Neurophysiology of Ingestion

Pergamon Studies in Neuroscience No 6

Series Editor Dr William Winlow,
Department of Physiology, University of Leeds, LS2 9NQ, UK

Neuroscience is one of the major growth areas in the biological sciences and draws both techniques and ideas from many other scientific disciplines. *Pergamon studies in neuroscience* presents both monographs and multi-author volumes drawn from the whole range of the subject and brings together the subdisciplines that have arisen from the recent explosive development of the neurosciences.

Pergamon studies in neuroscience includes contributions from molecular and cellular neurobiology, developmental neuroscience (including tissue culture), neural networks and systems research (both clinical and basic) and behavioural neuroscience (including ethology). The series is designed to appeal to research workers in clinical and basic neuroscience, their graduate students and advanced undergraduates with an interest in the subject.

1. **Function and dysfunction in the basal ganglia** *ed.* A.J. Franks, J.W. Ironside, R.H.S. Mindham, R.J. Smith, E.G.S. Spokes *and* W. Winlow

2. **Comparative aspects of neuropeptide function** *ed.* Ernst Florey *and* George B. Stefano

3. **Neuromuscular transmission: basic and applied aspects** *ed.* Angela Vincent *and* Dennis Way

4. **Neurobiology of motor programme selection** *ed.* Jenny Kien, Catherine R. McCrohan *and* William Winlow

5. **Interleukin-1 in the brain** *ed.* Nancy Rothwell *and* Robert Dantzer

New in 1993

A theory of the striatum   J. Wickens

Glycobiology and the brain   *ed.* M. Nicolini *and* P.F. Zatta

Neural modeling and neural networks   *ed.* F. Ventriglia

Neurophysiology of ingestion   *ed.* D.A. Booth

Neuroregulatory mechanisms in aging   *ed.* Maynard H. Makman *and* George B. Stefano

Thalamic networks for relay and modulation   *ed.* Diego Minciacchi, Marco Molinari, Giorgio Macchi *and* Edward G. Jones

Neurophysiology of Ingestion

Edited by

D. A. BOOTH

PERGAMON PRESS

OXFORD   •   NEW YORK   •   SEOUL   •   TOKYO

UK   Pergamon Press Ltd, Headington Hill Hall,
Oxford OX3 0BW, England

USA   Pergamon Press Inc., 660 White Plains Road,
Tarrytown, New York 10591-5153, USA

KOREA   Pergamon Press Korea, KPO Box 315, Seoul 110-603, Korea

JAPAN   Pergamon Press Japan, Tsunashima Building Annex,
3-20-12 Yushima, Bunkyo-ku, Tokyo 113, Japan

First edition 1993

Library of Congress Cataloging in Publication Data
A catalogue record for this book is available from the Library of Congress

British Library Cataloguing in Publication Data
A catalogue record for th is book is available from the British Library

ISBN 0 08 041988 7

Printed in Great Britain by B.P.C.C. Wheatons Ltd, Exeter

Contents

Preface

THIS book is centred on electrical recordings from single neurones that are related to the control of food or drink consumption. This is because of my conviction that unit recording is the key to understanding how the brain works.

There has been half a century of vigorous research into the neuroscience of intake. Obese or self-starving rats with damage in the hypothalamus have been famous among generations of students of psychology and neurobiology. Nevertheless, this book is the first to focus on bringing together the major contributions of integrative cellular neurophysiology to the understanding of the neural bases of mammalian eating and drinking behaviour. The reader is invited to consider how important a step in the right direction the appearance of this little volume might be.

The book is written for scientists and students who are interested in the brain mechanisms of eating and drinking, whether or not they work in neuroscience or ingestive behaviour. The chapter authors have been encouraged to review their own and others' research on a particular aspect of ingestive neuroscience in a way that is accessible to biological or behavioural scientists in other areas, but that can also serve to update colleagues working on other aspects of the neural bases of food and water intake and choice.

I am most grateful to my fellow authors for the enthusiasm, dedication and patience that they have shown during this enterprise. Also, without the invitation from Bill Winlow to contribute to the Pergamon Studies in Neuroscience series, this sort of book would still not have been produced, under my editorship at least.

Food and drink are fun. Research is a highly involving activity. Communication of ideas can bring great satisfaction when it materialises in print. I hope that readers too will share the pleasures of chewing over and digesting these issues about how we eat and drink, while regularly continuing to enjoy ingestive activity itself!

Edgbaston, July 1993   DAVID BOOTH

1

A framework for neurophysiological studies of ingestion

D. A. BOOTH

Neuroscience Laboratories, School of Psychology, University of Birmingham, England

THE neuroscience of ingestion is a long-established and vigorous area of research activity, represented in many of the textbooks. Intakes of food and drink became a topic for experiments on the brain in the 1940s, before electrical recording from nerve cells was generally feasible. In 1962, at the first conference on food and fluid intake as a satellite to the Congress of the International Union of Physiological Sciences, much of the work reported involved lesions or stimulation of rats' brains. Since 1970, Feeding and Drinking or, latterly, Ingestive Behaviour has been an area with its own series of sessions at the annual meetings of the Society for Neuroscience.

Yet research into the neural mechanisms of vertebrate food and water intake has not been obviously part of the mainstream of brain research. Invertebrate ingestion has been closer to other neuroscience but has had few links with intake by mammals and birds.

The 1940s also saw the beginning of theoretical work on the embodiment of cognitive functions in cellular interactions within a nervous system linked to its environment. This start on a mechanistic science of brain and behaviour was based on considerable neuroanatomical evidence and early findings from cellular electrophysiology. When the technique of unit recording became more widely applicable in the 1960s, systematic detail could be collected which has provided the foundation for a fully fledged science of integrative neurophysiology.

However, the psychologists and biologists of vertebrate intake rather seldom attempted to relate their research to evidence on what the nerve cells actually did. Neither theories nor data on the electrical activity of systems of neuronal units have been a major or central theme in the research journals or textbook

1

chapters on the physiology of hunger and thirst. Yet nobody doubts that interactions among cellular networks in the brain provide a key physical basis for eating and drinking behaviour.

There are now well-established techniques for single-cell recording within the brain of conscious and moving animals. A good number of neurophysiological laboratories have worked on particular aspects of systems involved in ingestive behaviour. Nevertheless, it remains a severe challenge to relate the results effectively to functionally integrative mechanisms. Also, this sort of work tends to be scattered in specialized research literature, either of general neurophysiology or of the sensory system or other area of physiology to which the results also relate. This book appears to be the first to bring together the major electrophysiological contributions to the understanding of ingestive behaviour. It provides ready access to the approach for scientists and students who are more familiar with other techniques for investigating brain mechanisms of eating and drinking. It should also facilitate exchange among the neuroscientists who specialize in separate aspects of the neural systems of ingestion.

1.1. The centrality of neurophysiology

This book focuses on cellular neurophysiology. That is because of the editor's longstanding conviction that recording the electrical activity of functionally and anatomically identified single neurones is the central foundation of neuroscience.

No amount of molecular neuroscience, cognitive neuropsychology or neural network theory, by themselves or together, can tell us how the central nervous system works to perform its functions. Cellular biochemistry cannot tell us how the brain works as a whole. Even the most detailed analysis of the behavioural effects of interfering with the brain by anatomically or chemically specific damage, blockade or stimulation fails to show how neurones are transforming information. The most refined forms of neuropsychology and psychopharmacology can only test theories of mechanisms when these have already been built from microanatomical and electrical records that reveal the actual connections between nerve cells.

The electrophysiological analyses also need to be of integrative neuronal interactions. Unless we specify the manner in which the firing patterns in one relevant sort of neurone interact with those of other neurones to influence the firing of another sort of cell in a functional subsystem, we cannot build a realistic theory of the neural processes underlying the organization of behaviour. Indeed, for the purpose of determining how the processes in a critical region of the brain contribute to the organization of behaviour and cognition, the other neurosciences are tools for testing the hypotheses generated by integrative neurophysiology.

Molecular neuroscience in itself can at most tell us what nerve cells do. Cognitive neuropsychology by itself tells us only about how the mind works,

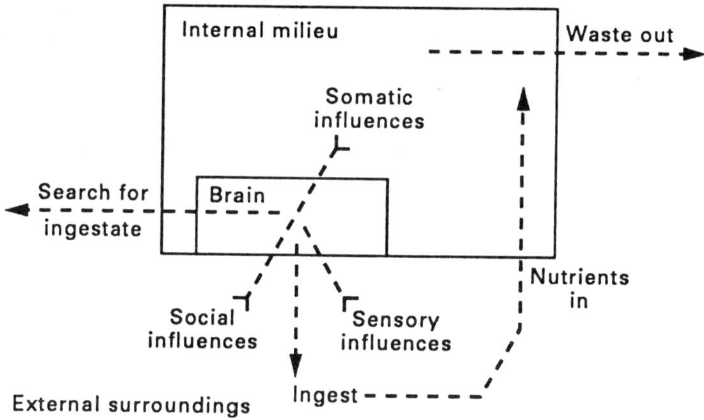

FIG. 1.1.   The input/output relationships constituting ingestive behaviour.

better than we can know from undamaged brains. Parallel distributed processing calculations alone can only tell us what is logically possible. Neurophysiological research is the only way to work out how nerve cells interact with each other and with the environment in network systems that perform the functions of the brain. Solving the age-old brain–mind problem will depend on integrative cellular neurophysiology related to cognitive analysis of behavioural performance.

In that spirit, the reader is invited to work through the specialized chapters that follow and catch a vision of what another half century of ingestive neuroscience could tell us about the brain and behaviour. Eating and drinking are not only fascinating in their own right. Their amenability to neurophysiological analysis makes them likely to be productive also of wider insights into the neural basis of behaviour.

1.2. Selection of appropriate ingestates

Ingestion is not just movements or amounts consumed. Ingestion is behaviour— that is to say, causal relationships between motor output and sensory input. Eating and drinking depend on the sensory control of reaching for the dietary material, its introduction into the mouth, chewing, transfer to the back of the mouth and swallowing (Fig. 1.1). This is evident from the fact that not every biteable solid or sippable fluid is ingested. Furthermore, even the materials that are not rejected on sight or smell, or spat out on tasting or feeling in the mouth, are selected, chewed and swallowed with varying degrees of apparent enthusiasm.

The intake of food and fluid therefore is not merely the disappearance of materials down the throat. Rather, each mouthful taken is the result of perceptual processing, both of sources of nutrition and of nutritional requirements of the body. The basic problem that needs explaining is how the searcher for food or drink knows when to eat or drink and what to ingest next, from information available in the external and internal environments. Investigating the control of each mouthful is closer to the timebase on which integrative neural processes operate than collecting data on amounts consumed in meals or over fixed periods.

Taking a drink, eating a food item and expressing dietary preferences all involve choices among materials according to their sensed characteristics. That is, items of food and drink must be discriminated and recognized. Hence, the technology of food and beverage formulation is indispensable to an effective sensory neurophysiology and psychology of eating and drinking. One of the most productive approaches to ingestive behaviour, predominant in this book, is the analysis of the oronasal senses involved.

In addition, however, choices among ingestates and decisions to start or stop consumption are at least to some degree coordinated to requirements for water, energy and other nutrients. This means that incipient states of need and repletion in the body must also be perceived in a way that enables appropriate action. Hence the visceral senses are also included.

In short, ingestion requires the recognition of items of food and drink, both from their sensory qualities and from predictable after-effects. This introductory chapter therefore outlines the problems that have to be solved by the eater's nervous system in recognizing foodstuffs and also in distinguishing bodily states in which it is appropriate to eat and to stop eating or indeed to take a drink (Booth, 1991).

1.3. Learning about foods

The detailed recognition of particular types of object and material by mammalian species is generally a learned performance. Some stimulus patterns are adequate to elicit behaviour innately. Yet even responses as reflexive as ingestion appears to be in normal and familiar situations are probably under control by patterns of stimulation that has been acquired by processes such as long-term habituation, associative conditioning and concept formation.

Both fixed reactions and instrumental acts towards food stimuli are conditionable in the infant rat within a few days of birth. Human babies can learn the smell, visual appearance and perhaps sound of the voice of their nursing mothers within a few weeks of birth at the most. Specifically to ingestion, there is evidence that human infants rapidly come to prefer the low level of sodium ions to which they have been exposed in breast milk, once the taste receptors have matured sufficiently for them to detect it (Harris, Thomas and Booth, 1991). Certainly, young rats that are beginning to feed spontaneously away from the dam learn to prefer whatever smell is given to energy- or protein-rich synthetic diets (Booth, Nicholls and Stoloff, 1974).

Human infants starting on solids learn to like salted foods and drinks if they are presented (Harris and Booth, 1985). Babies fed on sugar water still like plain sweetened water when they are 2 years old, unlike those who did not have that experience (Beauchamp and Moran, 1982).

These learned taste and odour preferences are not just for the sensory quality, however. They are specific to its intensity and also to the context of other sensory characteristics of the foodstuff. The older infant's preference is for the particular level of salt, for example, in the specific item of food or drink that has repeatedly been experienced with that level in it (Harris and Booth, 1987). Human adults show a sharp peak of preference among different salt levels in bread, close to the level generally marketed and experienced (Conner, Booth, Clifton and Griffiths, 1988). Similarly, in adult rats, the particular level of sweetness becomes preferred that has been associated with a higher level of osmotically mild carbohydrate calories (Booth, Lovett and McSherry, 1972).

1.3.1. *Sweetness preferences and aversions*

This calorically conditioned sweetness preference can completely overwhelm the unlearned liking for any sweetness (Booth *et al.*, 1972). This suppression of innate preference by learned preference is also evident in human adults. People come to like whichever level of sweetness is familiar in a particular food or drink. This level can differ widely between types of food. Yet, whatever the familiar level, samples of that food or drink containing appreciably higher levels of sweetness are less liked (independently of attitudes to sugars or artificial sweeteners) (Conner and Booth, 1988).

A newborn baby, by contrast, sucks more vigorously, consumes a larger amount and smiles more reliably the stronger is the sweetness placed on the tongue. This infantile reflex is also apparent in a child faced with an unfamiliar or disliked food that has been strongly sweetened. In adults too, when a familiar sweet drink is made very much sweeter than usual, the innate liking may break through despite the contrast with the level that has been learned to be preferred.

Also, the association of sweetness with effects of a poison or overdose of insulin will condition an aversion to sweetness (Lovett and Booth, 1970). At least with sufficiently moderate aversions, the most vigorous rejection is probably at the particular level of sweetness that was conditioned, while substantially higher and even perhaps lower concentrations are less aversive. Osmotic effects of hypertonic sugar solutions also condition aversions to hitherto attractive sweetness (Le Magnen, 1959; Booth *et al.*, 1972; Davis and Smith, 1990), evident as slowing of intake from the start of a test drink.

Hence, increasingly strong stimulation of sucrose-sensitive receptors and their afferents from the mouth is not necessarily transformed into increasing excitation of ingestatory movements. Depending on the learning history of the individual and on any other learned stimuli present, the response to a particular concentration of sucrose will be greater than to some higher or lower concentrations.

At which stage in the sensory and motor pathways this peak of recognition and preference is introduced by learning is a matter for neurophysiological investigation. It may differ between species and, within a species, between types of learning. For example, aversion conditioning alters the representation of sweetness at the first relay in the brainstem in the rat (Scott, Chapter 7). Yet the habituation-like type of satiation induced by exposure to sweetness is not represented in the monkey until the sensory pathway reaches association cortex (Rolls, Chapter 9).

1.3.2. *Sensory norms in learned preferences*

The learning-induced preference or aversion peak among sweetener concentrations is striking for the case of sweetness because we expect instead to see behaviour showing "the more the better". However, this decline in response on either side of the learned value is a general feature of acquired sensory control of behaviour, called by psychologists of animal learning "the intradimensional stimulus generalization decrement".

Consider a response that has been learned to a stimulus on a particular dimension (such as a taste or a colour) at a particular value (slightly sweet or bright red). If a test value is presented that is substantially different from the trained value, the response is liable to be less strong: it will not generalize well to that test stimulus. The less the test value is like the learned value, whether higher or lower on that dimension, the weaker the response will be. The response may fail to generalize at all to a much lower or much higher value on that dimension: that would be a complete generalization decrement. Values closer to the learned stimulus value result in only a partial decrement: the preference or the aversion is only somewhat reduced.

In other words, the effect of a stimulus on a response should be proportional to the perceptible difference between the test level and the learned level. Recognition of some aspect of a food (or indeed of an eating situation) as being at the preferred level can be regarded as an inability of the preference response to discriminate between those two levels.

The limit of discrimination between two levels of a stimulus is known in psychology as the "just noticeable difference" (JND) or the difference threshold (Torgerson, 1958). This is a measure of the strength of influence of an input over an output, however, for any process in any sort of system. There can be JNDs of stimuli affecting ratings of intensity or liking, preference measured as relative intakes, or neural activity in a sensory fibre or brain region. Furthermore, the JND is independent of the experimenter's units of measurement, unlike measures such as stimulus–response slope, discrimination errors or tuning bandwidth.

People's preferences for tastes, smells and textures in familiar foods and drinks do indeed decline on either side of the personally ideal level in proportion to the JND in the test circumstances (Fig. 1.2; Booth, Thompson and Shahedian,

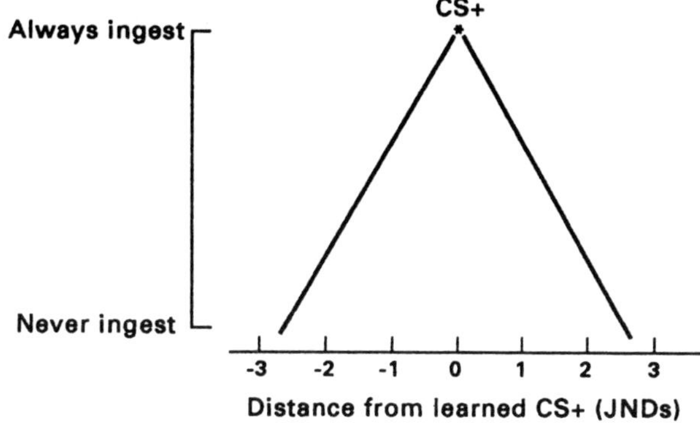

FIG. 1.2.   Decline in recognizability and preference with distance in JNDs of the test item or situation above or below the value of its learned form. This function becomes a straight line when the slope of the descending limb is multiplied by -1.

1983; Conner and Booth, 1992). Since the principle applies to visceral or external influences as well dietary ones, this measure of the strength of an influence on ingestive behaviour (and on ratings predictive of eating or drinking) has been called the "appetite triangle" (Booth, 1986). It represents the actual mechanism underlying the inverted U of the hedonic curves that are generated by averaging data across subjects, not plotting stimulus levels in JNDs and (as we shall see below) testing in unnatural situations.

This principle for measuring causal strength can be turned on its head to provide a technique for identifying the exact nature of an influence controlling a response. What the investigator initially assumes is the adequate stimulus may only approximate to the pattern on which the brain is actually working. If we can find a more accurate specification of the effective stimulus, then it will give a smaller JND. In other words, the intake, the rating or the neuronal firing rate will show sharper acuity. The stimulus measure to which the response is consistently most sensitive is thereby identified as the best candidate for being the actual source of influence (Kendal-Reed and Booth, 1992).

Such a search for the most effective stimulus can be used to help identify sensory receptor agonists. It can also specify a neural channel to which learned synaptic changes have tuned a pattern of stimulation to a set of diverse receptors, e.g. from a recognized food item or a whole situation motivating ingestion.

1.3.3. *Multiple controls of normal ingestion*

This approach also provides the means to identify how different pattern-signalling channels interact during the recognition of a multi-feature object, such as the mixture of foods suited to current bodily need and external situation. The pattern-combination rule that gives the lowest JND is the best candidate from

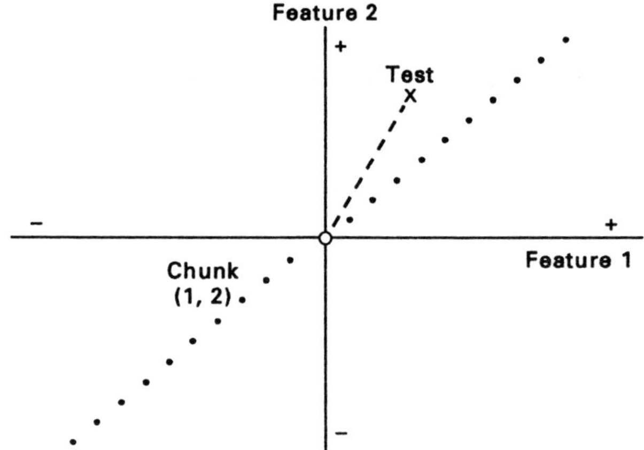

FIG. 1.3.   Two simple or complex features of an object or situation that can vary in values independently of each other. Origin: The normal values of the two features. X: test version.

that evidence for the actual form of the interaction among effects within the brain.

Signals over the same channel will add together in JND units. Different channels, however, are orthogonal: they form dimensions at right angles to each other. Recognition and preference therefore decline with the multidimensional distance of the test stimulus from the learned configuration (Fig. 1.3). This distance is the square root of the sum of the squares of the JND-scaled distances in each of the dimensions entering the decision to ingest (or not). Hence, the salient influences on ingestion may interact non-additively.

Furthermore, separate patterns can become one by a learning process of unreinforced association or "chunking" (Wickelgren, 1979). Two or more features, consistently present at particular levels in a learned food or eating situation, can emerge in overall appetite as a single integral dimension (the diagonal in Fig. 1.3). The learned response will decline as the strength of the unitary combination rises or falls from the learned value and as the features' strengths depart from their learned proportions (as in the test stimulus in Fig. 1.3).

For example, for some people a savoury taste of greater or less intensity than preferred in a vegetable- or meat-based food emerges from mixtures of sweetener, salt, acid and bitter in proportions that mimic the complex taste pattern of monosodium glutamate (Fig. 1.4; Booth, Freeman and Lähteenmäki, 1991). The overall taste is unique but nonetheless, for this person, it is no more than the JND-scaled Euclidean combination of other patterns of gustatory stimulation: there is no need to invoke another type of receptor. The salty tastes of sodium chloride and monosodium glutamate, on the other hand, add together in JND

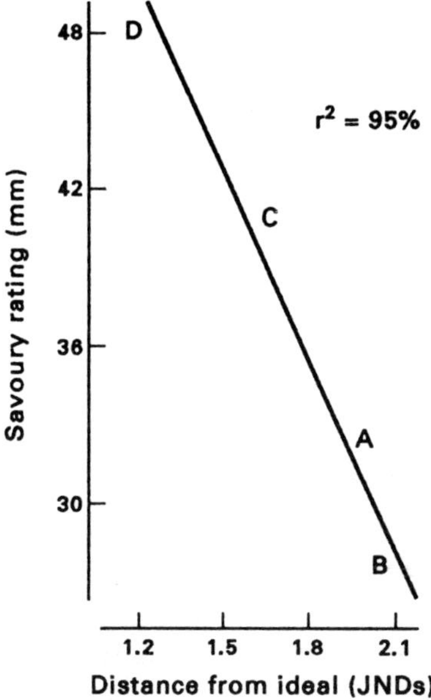

FIG. 1.4.   Multidimensional JND-scaled distances of four mixtures of sucrose, sodium chloride, citric acid and caffeine from their ideal levels for an individual, plotted against the difference from ideal "savoury" taste (50 mm) rated by that person on tasting a sample of each mixture in chicken soup.

units and so presumably result from action on the same sorts of sodium receptors.

Multidimensional discrimination analysis also provides a quantitative account of tests on poor-quality foods or in unusual visceral states or external situations. The distance of the conditions of testing from what the human being or other animal expects of these features will put an upper limit on the disposition to ingest. This defect will combine according to Pythagoras's Theorem with whatever is being experimentally varied within the animal's range of recognition. This results in a low and rounded but still mathematically determinate peak to the appetite triangle (Fig. 1.5). Tuning sensitivity can therefore still be measured in situations that are difficult to control completely, so long as they do not depart too far from what the organism knows well.

To sum up then, a person, monkey or rat faced with a familiar diet in a recognizable internal state and normal surroundings will respond according to the distance of the combination of salient features in the whole situation from a situation for which a response has been learned. If the response is sufficiently facilitatory, the animal will eat or drink. If it is not, or there is active inhibition, the animal will not ingest.

What the individual ingests will depend on the relative contributions of the sensory qualities of the different foodstuffs present to overall facilitation. This

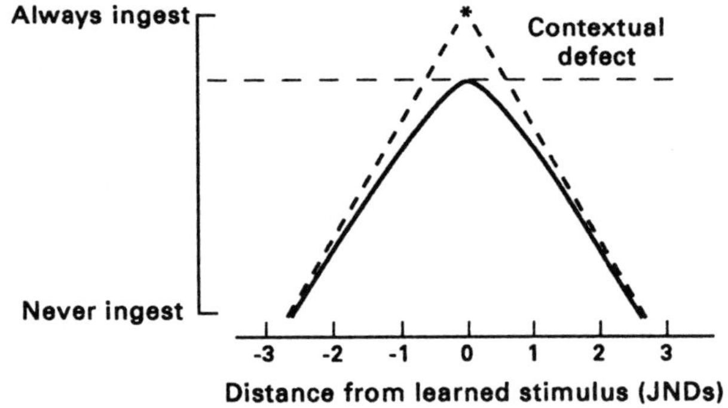

FIG. 1.5.   Preference (or difference from familiar) for test samples varying in a simple or complex feature and having a constant defect (also uni- or multidimensional). When the distance of the varied feature from the learned level is sufficiently greater than the defect, this function approximates to the isosceles triangle of Fig. 1.2.

immediate food-generated facilitation is generally called *preference*. The preference for a food might be called its *palatability* if sensory preference is assumed not to depend on circumstances, but that is not generally true. Whether the individual ingests may also depends on visceral facilitation: this effect is often called "hunger" but that term has a wider sense of appetite for food (as opposed to "thirst" which is water appetite), whether facilitated internally or externally. Ingestion is also contingent on factors such as risk of predation in small mammals, the cost of effort (or, in people, of money) and social context (e.g. which other conspecifics are eating). This is sometimes called the *appropriateness* of the occasion, but that may also include the goodness of match of a food to other items on the menu and even to bodily state.

1.3.4. *Control of ingestion by its after-effects*

When ingestive inhibition arises from the effects of recent eating, this is known as a state of *satiety*. Satiating effects do not all arise from postingestional actions of food and drink, such as stretch of the stomach, chemical stimulation of the upper intestine or hepatic metabolism. Eating may stop because a prefilled plate is empty. Drinking may end because the contents of the cup or can are finished. The eater may simply have had enough of the taste of a food for a while, independently of how full the gut is.

Repeated exposure to a stimulus pattern causes the initial orienting, ingestive or defensive response to habituate. The response recovers after some minutes. Also, the introduction of a new stimulus (variety) can still excite the response, at least until continued presentation induces habituation to that stimulus also. When series of presentations are repeated, the habituation becomes faster. This shows that some residual memory of the previously habituated stimulus remains and is called long-term habituation. It may be one of several mechanisms in food-specific satiation. Satiation that arises from habituation rather than from learning about the after-effects of food consumption cannot contribute to regulation of energy balance through control of amount eaten.

Unvarying conditioned aversions and preferences cannot adjust the amount consumed to its caloric yield either, because these effects operate from the start of the meal, independently of how much is eaten.

The learned control of meal size on a particular food will be related to caloric yield only if an aversion (or at the least a lack of preference) can be confined to the end of a meal and controlled by how much energy has been ingested as well as by the food's sensory characteristics. Thus, calorically regulatory satiation at a meal requires a loss of ingestive facilitation that is specific both to a food and to a level of gastrointestinal repletion that have together been followed by an overabundance of dietary energy (Booth, 1972b; Booth and Davis, 1973; Gibson and Booth, 1989).

Besides this inhibitory learning, many natural associates of eating condition facilitation to stimulus combinations that predict them. These include postingestional effects of carbohydrate, fat and alcohol, perhaps the fuel they provide to the liver (Booth, 1972a). Other nutrients such as the essential amino acids from dietary protein can have the same ingestion-facilitating effects. All these nutrients can condition preferences to simple or complex food tastes, aromas, tactile textures and visual appearances. They also condition ingestion to combinations of food characteristics and bodily states, such as a full or empty gut (Booth and Davis, 1973; Gibson and Booth, 1979) or an incipient lack of essential amino acids (Gibson and Booth, 1986a; Booth and Baker, 1990). These provide qualitative (categorical) examples of the two-dimensional integration represented quantitatively in Fig. 1.3. Such mechanisms help to improve the match between nutrient requirements and food selection and intakes, although they do not provide upper limits on the intake of appetizing foods in hungrifying states.

Social and/or emotional associates also associatively condition appetite or its reduction. Novel foods given to young children with some praise become more attractive (Birch, 1990). Conversely, rewarding the eating of a food with a liked food reduces preference for the first food (Birch, 1990). External signals can also be conditioned to elicit eating (Weingarten, 1984). Thus, the social situations or emotional states that reinforce ingestion could conceivably also become parts of the learned complexes of stimuli for ingestive appetite. This would be a mechanism for emotional eating which could override learned visceral control of preferences and satiety.

1.4. Neurophysiologically oriented ingestive neuroscience

The work represented in this book is relevant to any research into neural mechanisms of ingestion, even if electrical recordings are not being made. Studies of intake and the brain should be designed to relate to what is known from integrative neurophysiology, as well as being sensorially and viscerally contextualized. A few examples are now outlined of such combinations of behavioural analysis, somatic control and neurophysiological ideas in the study of the central pathways of ingestion by other methods (Booth, 1990).

1.4.1. *Are there centres for appetite and satiety?*

Before the rise of cellular neurophysiology, the main approach in what was called physiological psychology was to analyse the behavioural consequences of destroying certain regions of the brain; this approach is now named neuropsychology. Lesions of the ventromedial region of the hypothalamus (VMH) made rats overeat and get very fat, while lesions in the far lateral hypothalamus (LH) stopped eating altogether or at least for some weeks. So a dual-centre theory of hunger and satiety was proposed, that the LH was the location at which feeding was organized and the VMH was the satiety centre. This idea seemed to be strongly supported by evidence that electrical stimulation of the LH could trigger eating and VMH stimulation could stop it, while activity recorded through large electrodes in the LH and VMH varied reciprocally in food-deprived and fed rats.

A similar pattern of evidence was the basis for proposing initially that the medulla contained dual centres for breathing in and breathing out. However, such a convergence of effects of lesions, stimulation and recording does not establish the location of an organizing centre, let alone its existence. For example, there might well be no more than efferent axons in such a region (Booth, 1968b). It was soon shown that the supposed medullary "centres" are indeed largely efferent pathways and expiration and inspiration are organized in complex networks extending through the pons into the midbrain (Porter, 1970). There has long been evidence for the importance of extrahypothalamic mechanisms in food intake. The other chapters in this book illustrate recent findings of that sort.

In any case, the postulation of a centre (even if it were confirmed to exist) would do nothing to explain either the neural mechanisms or the behavioural processes by which the function is achieved. It just transfers the "black box" from a boundary around the brain or body to a name for a place inside the brain. This strategy was called phrenology in the nineteenth century: it is no less fallacious if based on drug action or electrical recordings instead of on the effects of lesions (Booth, 1976). Scientific advance in integrative neuroscience requires the coordination of neurophysiology and microanatomy into a theory of neuronal networks that could perform a function such as organizing ingestive behaviour (Rolls, 1976, and in the final chapter of this book).

By the early 1970s, further analysis of effects of manipulating either the LH or the VMH had completely refuted the dual-centre theory of hunger and its satiety (Rabin, 1972; Ungerstedt, 1971). Yet this situation has still not permeated through to textbooks nor to some other areas of research in neuroscience. A selection of contrary evidence is now summarized. These are not neurophysiological data but they serve to illustrate how the above integrative behavioural framework can be used to approach the subsequent chapters that do centre on electrical recording and other evidence of cellular interconnections.

1.4.2. Hypothalamic synaptic fields in eating control

The first evidence against a lateral hypothalamic hunger centre came from improved anatomical specification of the elicitation of eating in the sated rat by injection of noradrenaline into the hypothalamus (Booth, 1967, 1968a). The microcannulae developed have not been improved on but they cannot imply a site of action with a precision of more than about 0.2 mm. Nevertheless, the noradrenaline feeding site was clearly the best part of a millimetre anterior to the lateral hypothalamic sites at which electrical stimulation was most likely to elicit eating, at the level where lesions most reliably caused aphagia (Booth, 1967).

Subsequent work focusing on the most effective region for adrenergic elicitation of eating identified a small area in or just lateral to the paraventricular nucleus, which is well rostral and somewhat medial to the LH (Liebowitz, 1978; Matthews, Booth and Stolerman, 1978). Rolls (1976 and this volume) found eating-related units in the lateral hypothalamus of monkeys but also more dorsal in the substantia innominata.

1.4.3. Ventromedial hypothalamic obesity

Experimental lesions of the hypothalamus (and tumours there) alter the autonomic balance at the stomach, endocrine pancreas, liver and white and brown fat, shifting defended body weight and sometimes altering eating in the process. Lesions in the ventromedial hypothalamic area immediately release gastric emptying from inhibition that it is normally under, especially during the rat's inactive period in the first part of daylight (Duggan and Booth, 1986, 1991). The resulting faster digestion and absorption stimulates extra insulin secretion and fat deposition, sufficiently to increase food intake on a palatable diet, but in any case making the rat more and more obese until feedback on satiety from the fat stores turns the dynamic phase into an asymptotic static phase of obesity (Toates and Booth, 1974). The continuous rapid absorption caused by the accelerated stomach soon adapts the autonomic balance and transmitter sensitivities on the pancreatic beta cell (if the ventromedial hypothalamic lesion has not already shifted the balance) and even more insulin is released, producing still more fat deposition (Campfield, Smith and Fung, 1982).

These changes are allied with normal central mechanisms controlling eating behaviour and its responsiveness to visceral signals. There are no defects in postingestional satiety mechanisms in rats with ventromedial hypothalamic lesions (Booth, Toates and Platt, 1976; Duggan and Booth, 1986). So the ventromedial nucleus and its surrounding hypothalamic region cannot be serving as a satiety centre. The effects of signals from the stomach and from circulating substances recorded in the VMH nucleus presumably act on neuroendocrine and autonomic outflows. Any effect on visceral inhibition of eating is generated secondarily in the periphery. For example, rapid gastric emptying will end nutrient absorption early, prematurely removing stimulation from intestinal chemoreceptors (Welch, Sepple and Read, 1988) and so releasing frequent meals (Booth, 1978).

1.4.4. Behavioural analysis of adrenergic eating

The brain organizes particular externally observable processes that operate over time periods from tens of milliseconds to seconds at the most. Thus, little of the information needed to build a neuroscience of ingestion can be extracted from measures of the accumulated consequences of much changing ingestive behaviour, like the amount eaten or drunk in 30 minutes or in a meal. This chapter therefore introduces forms of behavioural analysis that deal with time periods short enough for input/output relations not to change. The data can then be related to neural processes specified by electrical recording from single cells.

The adrenergic eating effect from the rostral hypothalamus has been approached this way. Since it was discovered that infusion of noradrenaline into the sensitive region would increase the size of an ongoing meal (Ritter and Epstein, 1975), as well as elicit eating, it has been thought that the effect was disruption of satiety. Behavioural analysis provided clear evidence against an increase in learned or unlearned sensory preferences (Matthews, Gibson and Booth, 1985) or indeed a decrease in aversiveness of the test diet (Sclafani and Toris, 1981).

Nevertheless, an injection evoking ingestion of sweetened condensed milk from rats sated half an hour previously did not elicit milk ingestion immediately after gastric intubation of the amount that the rats consumed after the usual presatiation procedure (Gibson and Booth, 1986b). This showed that the behavioural effect of noradrenaline did not depend on the blockade of all satiating influences, certainly not of strong postingestional satieties.

The only satiety effect found to be overcome by the injection was conditioned satiation, the learned decrease in flavour preference when the stomach was partly full (Matthews *et al.*, 1985; Booth, Gibson and Baker, 1987). The control of ingestion by the learned combination of dietary and visceral stimuli can be rather strong (Booth, 1985). Nevertheless, it requires complex processes of crossmodal integration with memory. Therefore, it is liable to be susceptible to cognitively disrupting factors such as excessive arousal. The noradrenergic synapses on which the injection acts are a local projection of the ascending reticular arousal systems. One may therefore wonder whether this is one of the neural mechanisms involved in emotional overeating.

1.5. The structure of this book

Research progress can only be made by focusing on technically answerable questions. This often requires specialization. Neurophysiological investigations of ingestive behaviour or the mechanisms controlling food or water intakes have generally focused on a particular category of afferent information and on the efferent modulation over which those inputs are predominant.

A book has to be a sequence of chapters. Yet we have seen that ingestion is controlled by conjunctions of afferent activities rather than by a sequence of signal onsets or offsets. Similarly, the movements of the mouthparts and of food handling during ingestion are largely co-temporaneous, even though obviously there are also sequences of approach and consumption and cycles of biting, chewing and swallowing.

To emphasize, therefore, the visceral context within which any ingestion or refusal of food or drink takes place, the first four specialized chapters deal with the intake-related neurophysiology of the liver, intestine and stomach. Then we consider the tactile control of the movements of the jaw and tongue that are involved in all eating and drinking. Next come two of the other oral senses, taste and smell (which is retronasal for the aroma of food in the mouth as well as orthonasal for sniffed food odours). Finally, we consider examples of cross-modal integration between the appearance, taste and smell of a food and the neural systems transforming recognition of a food into action towards it.

The authors have been encouraged to illustrate their chapters by neurophysiological findings from their own laboratories while providing a brief overview of that aspect of the neurophysiology of ingestion for the non-expert. There is not the space nor perhaps are we yet at the time to bring out the connections and the gaps between the different lines of research. This is the first collection of reviews for a wide readership that covers virtually the whole range of neurophysiological approaches to ingestive behaviour in mammals. As such it can serve as something of a landmark in research activity within behavioural neuroscience as well as an introduction for newcomers to this intriguing topic for investigation.

References

Beauchamp, G. K. and Moran, M. (1982). Dietary experience and sweet taste preferences in human infants. *Appetite*, **3**, 139–152.

Birch, L. L. (1990). The control of food intake by young children: the role of learning. In E. D. Capaldi and T. L. Powley (Eds.), *Taste, Experience, and Feeding* (pp. 116–135). Washington DC: American Psychological Association.

Booth, D. A. (1967). Localization of the adrenergic feeding system in the rat diencephalon. *Science*, **158**, 515–517.

Booth, D. A. (1968a). Mechanism of action of norepinephrine in eliciting an eating response on injection into the rat hypothalamus. *Journal of Pharmacology and Experimental Therapeutics*, **160**, 336–348.

Booth, D. A. (1968b). Effects of intrahypothalamic glucose injection on eating and drinking elicited by insulin. *Journal of Comparative and Physiological Psychology* **65**, 13–16.

Booth, D. A. (1972a). Postabsorptively induced suppression of appetite and the energostatic control of feeding. *Physiology and Behavior*, **9**, 199–202.

Booth, D. A. (1972b). Conditioned satiety in the rat. *Journal of Comparative and Physiological Psychology*, **81**, 457–471.

Booth, D. A. (1976). Approaches to feeding control. In T. Silverstone (Ed.), *Appetite and Food Intake* (pp. 417–478). Berlin: Dahlem Konferenzen.

Booth, D. A. (1978). Prediction of feeding behaviour from energy flows in the rat. In D. A. Booth (Ed.), *Hunger Models: Computable Theory of Feeding Control* (pp. 227–278). London: Academic Press.

Booth, D. A. (1985). Food-conditioned eating preferences and aversions with interoceptive elements: learned appetites and satieties. *Annals of New York Academy of Sciences*, **443**, 22–37.

Booth, D. A. (1986). Objective measurement of influences on food choice. *Appetite*, **7**, 236–237.

Booth, D. A. (1990). The behavioral and neural sciences of ingestion. In E. M. Stricker (Ed.), *Neurobiology of Food and Fluid Intake* (pp. 465–488). New York: Plenum Press.

Booth, D. A. (1991). Influences on human drinking behaviour. In D. J. Ramsay and D. A. Booth (Eds.), *Thirst: Physiological and Psychological Aspects* (pp. 52–72). London: Springer.

Booth, D. A. and Baker, B. J. (1990). dl-Fenfluramine challenge to nutrient-specific textural preference conditioned by concurrent presentation of two diets. *Behavioral Neuroscience*, **104**, 226–229.

Booth, D. A. and Davis, J. D. (1973). Gastrointestinal factors in the acquisition of oral sensory control of satiation. *Physiology and Behavior*, **11**, 23–29.

Booth, D. A., Freeman, R. P. J. and Lähteenmäki, L. (1991). Likings for complex foods and meals. *Appetite* **17**, 156.

Booth, D. A., Gibson, E. L. and Baker, B. J. (1986). Behavioral dissection of the intake and dietary selection effects of injection of fenfluramine, amphetamine or PVN norepinephrine. *Society for Neuroscience Abstracts*, **15**, 593.

Booth, D. A., Lovett, D. and McSherry, G. M. (1972). Postingestive modulation of the sweetness preference gradient in the rat. *Journal of Comparative & Physiological Psychology*, **78**, 485–512.

Booth, D. A., Stoloff, R. and Nicholls, J. (1974). Dietary flavor acceptance in infant rats established by association with effects of nutrient composition. *Physiological Psychology*, **2**, 313–319.

Booth, D. A., Thompson, A. L. and Shahedian, B. (1983). A robust, brief measure of an individual's most preferred level of salt in an ordinary foodstuff. *Appetite*, **4**, 301–312.

Booth, D. A., Toates, F. M. and Platt, S. V. (1976). Control system for hunger and its implications in animals and man. In D. Novin, W. Wyrwicka and G. A. Bray (Eds.), *Hunger: Basic Mechanisms and Clinical Implications* (pp. 127–142). New York: Raven Press.

Campfield, L. A., Smith, F. J. and Fung, K. F. (1982). VMH hyperphagia and obesity. Role of autonomic neural control of insulin secretion. In B. G. Hoebel and D. Novin (Eds.), *Neural Basis of Feeding and Reward* (pp. 203–220). Brunswick ME: Haer Institute.

Conner, M. T. and Booth, D. A. (1988). Preferred sweetness of a lime drink and preference for sweet over non-sweet foods, related to sex and reported age and body weight. *Appetite*, **10**, 25–35.

Conner, M. T. and Booth, D. A. (1992). Combining measurement of food taste and consumer preference in the individual: reliability, precision and stability data. *Journal of Food Quality*, **15**, 1–17.

Conner, M. T., Booth, D. A., Clifton, V. J. and Griffiths, R. P. (1988). Individualized optimization of the salt content of white bread for acceptability. *Journal of Food Science*, **53**, 549–554.

Davis, J. D. and Smith, G. P. (1990). Learning to sham feed: behavioral adjustments to the absence of gastrointestinal stimulation. *American Journal of Physiology*, **259**, R1228–R1235.

Duggan, J. P. and Booth, D. A. (1986). Obesity, overeating and rapid gastric emptying in rats with ventromedial hypothalamic lesions. *Science*, **231**, 609–611.

Duggan, J. P. and Booth, D. A. (1991). Failure to demonstrate that accelerated gastric emptying after VMH lesions is secondary to excess weight gain. *American Journal of Physiology*, **261**, R515–R516.

Gibson, E. L. and Booth, D. A. (1986a). Acquired protein appetite in rats: dependence on a protein-specific need state. *Experientia*, **42**, 1003–1004.

Gibson, E. L. and Booth, D. A. (1986b). Feeding induced by injection of norepinephrine near the paraventricular nucleus is suppressed specifically by the early stages of strong postingestional satiety in the rat. *Physiological Psychology*, **14**, 98–103.

Gibson, E. L. and Booth, D. A. (1989). Dependence of carbohydrate-conditioned flavor preference on internal state in rats. *Learning and Motivation*, **20**, 36–47.

Harris, G. and Booth, D. A. (1985). Sodium preference in food and previous dietary experience in 6-month-old infants. *IRCS Medical Science*, **13**, 1177–1178.

Harris, G. and Booth, D. A. (1987). Infants' preference for salt in food: its dependence upon recent dietary experience. *Journal of Reproductive and Infant Psychology*, **5**, 97–104.

Harris, G., Thomas, A. and Booth, D. A. (1991). Development of salt taste preference in infancy. *Developmental Psychology*, **26**, 534–538.

Kendal-Reed, M. and Booth, D. A. (1992). Human odor perception by multidimensional discrimination from remembered patterns. Abstracts of 14th Annual Meeting of American Chemoreception Society, Saratoga, April 1992.

Leibowitz, S. F. (1978). Paraventricular nucleus: a primary site mediating adrenergic stimulation of feeding and drinking. *Pharmacology Biochemistry and Behavior*, **8**, 163–175.

Le Magnen, J. (1959). Effets des administrations postprandiales de glucose sur l'établissement des appétits. *Comptes Rendues des Séances de la Société de Biologie, Paris*, **153**, 212–215.

Lovett, D. and Booth, D. A. (1970). Four effects of exogenous insulin on food intake. *Quarterly Journal of Experimental Psychology*, **22**, 406–419.

Matthews, J. W., Booth, D. A. and Stolerman, I. P. (1978). Factors influencing feeding elicited by intracranial noradrenaline in rats. *Brain Research*, **141**, 119–128.

Matthews, J. W., Gibson, E. L. and Booth, D. A. (1985). Norepinephrine-facilitated eating: reduction in saccharin preference and conditioned flavor preferences with increase in quinine aversion. *Pharmacology Biochemistry and Behavior*, **22**, 1045–1052.

Porter, R. (Ed.) (1970). *Breathing*. London: CIBA.

Rabin, B. M. (1972). Ventromedial hypothalamic control of food intake and satiety—a reappraisal. *Brain Research*, **43**, 317–323.

Ritter, R. C. and Epstein, A. N. (1975). Control of meal size by central noradrenergic action. *Proceedings of National Academy of Sciences, U.S.A.*, **75**, 3740–3743.

Rolls, E. T. (1976). The neurophysiology of appetite. In T. Silverstone (Ed.), *Appetite and Food Intake* (pp. 21–42). Berlin: Dahlem Konferenzen.

Sclafani, A. and Toris, J. (1981). Influence of diet palatability on the noradrenergic feeding response in the rat. *Pharmacology Biochemistry and Behavior*, **15**, 15–19.

Toates, F. M. and Booth, D. A. (1974). Control of food intake by energy supply. *Nature*, **251**, 710–711.

Torgerson, W. S. (1958). *Theory and Methods of Scaling*. New York: Wiley.

Ungerstedt, U. (1971). Adipsia and aphagia after 6-hydroxydopamine induced degeneration of the nigro-striatal dopamine system. *Acta Physiologica Scandinavica*, **367**, (Supplement), 95–122.

Weingarten, H. P. (1984). Meal initiation controlled by learned cues: basic behavioral properties. *Appetite*, **5**, 147–158.

Welch, I., Sepple, C. W. and Read, N. W. (1988). Comparison of the effects on satiety and eating behaviour by infusion of lipid emulsion into different regions of the small intestine. *Gut*, **29**, 306–311.

Wickelgren, W. A. (1979). Chunking and consolidation: a theoretical synthesis. *Psychological Review*, **86**, 44–60.

2

Regulatory control of food and water intake and metabolism by the liver*

DONALD NOVIN

Psychology Department and the Brain Research Institute, University of California, Los Angeles, CA 90024, USA

THE importance of the liver in the regulation of substrates for energy and maintenance of caloric homeostasis was well established in the first third of this century (Cahill, 1977). It is therefore surprising that a theory of hepatic control of food intake was only first proposed by Russek in 1963. A neurophysiology of the liver did not seem relevant while the hypothalamic dual-centre concept held sway from the fifties to near contemporary times (Novin and VanderWeele, 1977).

This chapter argues a critical role for the liver in the control of ingestion. However, in no way does it argue against a role for other visceral organs and systems in caloric regulation (see Novin, 1988, for a more detailed discussion of that issue). For a deeper understanding of hunger and satiety mechanisms, it is time to emphasize synthesis more than analysis. For example, how might the stomach and liver interact in the task of regulating energy metabolism through feeding?

2.1. Functional considerations

The "logic" of the liver makes this organ the prime candidate for regulatory control of food and water intake.

First, the fact that the liver is the principal organ in metabolic regulation makes it reasonable to seek hepatic control of feeding, linking this behaviour to availability of energy to the tissues. This is the basis for the "energostatic" theory

---

*The research herein reported was largely supported by grants from the Academic Senate of UCLA.

of Booth (1972), and similar concepts of Friedman (1991) and Scharrer and Langhans (1988).

Secondly, the availability of energy substrates to tissues and the overall utilization of calories is nearly unchanged when an overnight fast is extended to 3 days of food deprivation in an adult vertebrate. Energy metabolism of the brain is especially well maintained, which would argue against the importance of central nervous system receptors. In contrast, the metabolism of the liver is drastically changed within a few hours of food deprivation. In freely feeding mammals, 4 hours after a meal, the liver changes from an organ with a net uptake of glucose to one of output. The teleology of this is, of course, to ensure constancy of energy supply for the vital organs. Thus, monitors of hepatic metabolism could "tell" the brain much of the information needed to control food intake accurately. In contrast, the brain is spared these net changes and so would not be the "logical" place to detect them. The brain is the beneficiary of this regulation, not the primary initiator.

There is electrophysiological evidence, thirdly, that at least some of the receptors needed to monitor the availability of energy and some individual nutrients (e.g. hexose sugars) can be found in the liver and their presence can alter afferent neural activity (Niijima, 1969). There is some question about their role in behaviour, however, as they might act entirely at the autonomic level (Niijima, 1981).

Finally, the liver is the first organ "downstream" to the digestive tract and inside the body in the topological sense. This makes it, then, the first organ to be able both to take energy-bearing substances from the circulation (through the capillary bed of hepatic-portal system) and also to be able to measure that uptake. This is another reason to consider it likely that the liver is involved in the control of hunger and satiety. However, studies following hepatic-portal bypass (Louis-Sylvester, 1981), surgically removing the direct connection between the intestinal and hepatic capillary beds, have shown little effect on the patterns of food intake such as meal sizes, inter-meal intervals and their interrelations. Thus, if the privileged position of the liver with respect to newly absorbed nutrients is an important factor in the regulatory control of feeding, more direct or refined measures of its role are needed.

2.2. Innervation of the liver

Most likely, the liver regulates metabolism through food intake by sensing some aspect of energy or nutrient availability. There is no clear evidence for any humoral agent other than energy substrates that could carry information from liver to brain. Therefore, it is likely that this information goes via the two parts of the autonomic nervous system (ANS), the parasympathetic (PNS) and the sympathetic (SNS) nervous systems, and there is considerable evidence for the proposition.

2.2.1. *Efferents into the liver*

The early descriptions of the ANS were primarily, if not exclusively, of an efferent or motor system. The efferent SNS and PNS are thought to act largely in opposition to one another. This antagonistic relation is clearly seen in the liver. Activation of the SNS efferent innervation of the liver, which is via the splanchnic nerves, causes a rise in blood glucose levels. This is due in part to an inhibition of insulin release and increased glucagon secretion by collateral ANS efferents to the endocrine pancreas. This hormonal pattern retards glucose uptake by most peripheral tissues, inhibits glycogen formation and activates glycogenolysis in the liver, thus releasing glucose into the blood stream. PNS activation yields an endocrine profile the reverse of that following SNS activity, resulting in an opposite picture of hepatic metabolism.

Shimazu (1967) showed that these same metabolic consequences also result from innervation of the hepatocytes by the two divisions of the ANS. Vagal PNS efferent fibres providing cholinergic innervation go largely through the hepatic branch. The only synaptic connections that these fibres make (Prechtl and Powley, 1985) is in the liver itself (and vagal paraganglia cells). It is likely that the liver receives some PNS innervation by nerve fibres carried within the largely SNS splanchnic nerves. SNS fibres in the splanchic nerves have synapses in the coeliac and mesenteric ganglia and travel with the fine arteries of the circulatory system and in the wall of the hepatic-portal vein and hepatic artery.

The cell bodies of the vagal efferent system are found mostly in the dorsal motor nucleus of the vagus (DMNV) and to a lesser extent in nucleus ambiguus of the brain stem. The cell bodies of the SNS efferent fibres lie primarily in the interomedio-lateral columns of the spinal cord.

2.2.2. *Afferents from the liver*

Both divisions of the ANS carry afferent as well as efferent fibres. The abdominal vagii of both cat and rat are made up of 80% or more sensory fibres. The cell bodies of the vagal PNS going to the liver and other visceral organs are located in the nodose ganglia. These bipolar cells give off one branch to form the vagus in the periphery. The other branch goes centrally to make synaptic contact with the neurones of the nucleus tractus solitarius (NTS) and to a lesser extent with the adjacent area postrema and the DMNV. The afferents in the SNS side of the ANS are not well known. Some of these fibres may travel with the vagus nerve to the NTS.

One peculiarity of the afferent projection of the liver is its bilateral asymmetry: virtually all of its representation is in the left NTS. This was first seen by Adachi (1981) using electrophysiological techniques and later confirmed by Rogers and Herman (1983) using horseradish peroxidase retrograde labelling and electrophysiology. The functional significance of this, if any, is not known.

2.3. Signals in the control of feeding

The most completely developed theory of hepatic control of food intake is one proposed by Scharrer and Langhans (1988). The principal metabolic control of hunger and satiety in this theory is the rate of respiratory chain activity or oxidative phosphorylation in hepatocyte mitochondria. Their evidence comes from a series of experiments where they looked at the effects on food intake of the administration of redox pairs. These are pairs of metabolites in which A can be oxidized to B and B can be reduced to A. (The metabolic path from A to B does not necessarily have to be the reverse of the B to A pathway.) They find that it is only the substance that is reduced and thus a potential substrate for oxidation that can suppress feeding. The oxidized member of the pair has little effect on feeding. It is known that the electron transfer resulting from oxidative phosphorylation produces energy in the form of terminal phosphate bonding in adenosine triphosphate (ATP). The increase in ATP levels increases the activity of the ATPase working the sodium/potassium pump in the cell membrane. This increased activity increases the membrane potential (hyperpolarizes the hepatocyte membrane) which would inhibit the activity of vagal afferents innervating those cells. Thus, increased availability of glucose (and possibly any other energy substrate that the liver can utilize) would reduce vagal activity and could thereby send a signal to the brain to reduce hunger.

While the scheme is speculative, there are nevertheless convincing data for each step of the whole process. Niijima (1983) showed that ouabain, which inhibits Na/K pump activity, significantly reduced the effects that hepatic portal glucose infusions had on hepatic vagal activity. Furthermore, Fitz and Scharschmidt (1987) showed that the membrane potential of liver cells recorded in fasted animals averaged 10 mV less (i.e. depolarized) than that of non-fasted rats.

Scharrer and Langhans (1988) themselves provided some of the most direct evidence in support of an hepatic control of feeding. Their experiments showed that increased mitochondrial oxidation in the liver decreased feeding, and blocking fatty acid oxidation increased feeding. Moreover, both of these effects depended upon an intact hepatic vagus nerve. This branch of the vagus is the only, or at least the principal, parasympathetic innervation of the liver. Its loss should substantially alter feeding behaviour if their theory were correct. With respect to these manipulations of hepatic energy metabolism, vagotomy clearly does just this.

Further evidence comes from observations that infusion of 2DG is more effective in stimulating feeding when given into the hepatic portal vein than when given via a jugular-vein cannula (Novin, VanderWeele and Rezek, 1973). This suggests that liver receptors are more involved than the brain in activating feeding mechanisms. Stricker and colleagues (1977) looked at the effects of different energy substrates on insulin-induced hypoglycaemia which causes animals to increase food intake and activates the SNS. They used ketones, fructose and glucose, which are utilized to different extents in the brain and liver. They found that glucose, which is utilized equally well by the brain and liver, was capable of reducing feeding as well as SNS discharge following hypoglycaemia. Ketones, which can get into and be utilized by the brain but not the liver, suppressed the SNS discharge but not the feeding. Fructose administration had the opposite effect to ketones because this sugar does not get across the blood–brain barrier and so circulating fructose is not therefore utilized by the brain. It enters the liver, however, and is rapidly oxidized there. Their conclusion was that feeding, at least to insulin hypoglycaemia, is generated by hepatic receptors more than by effects within the brain, whereas the SNS response is brain-driven and not evoked by an hepatic mechanism.

Another important theory about the signals that control food intake is the glucostatic hypothesis, which is that the utilization specifically of glucose from the blood suppresses hunger and a lack of glucose metabolism in the critical cells releases hunger. Electrophysiological work clearly supports the existence of glucose-detecting systems. For example, Niijima (1969) showed that there are a number of individual fibres in the hepatic vagus of the guinea pig which responded selectively to changes in hepatic-portal glucose concentration. The code was very simple: vagal neural activity was a linear but negative function of glucose concentration. This sort of evidence was thought by some to have been subsumed under the energy-regulatory theory of hepatic oxidation. However, Louis-Sylvestre and Le Magnen (1980) and Campfield, Brandon and Smith (1985) have shown perfect correlations between the occurrence of small hypoglycaemic episodes and feeding bouts in rats and therefore suggest some special role for glucose as distinct from the calories that it contains. Some advocates of hepatic oxidation have in fact allowed that this is an additional possibility (Booth, Toates and Platt, 1976; Novin, O'Farrell, Acevedo-Cruz and Geiselman, 1991). These glucostatic signals are thought also to be detected by the liver, on the basis of results from a variety of denervation studies (for a summary, see Novin, 1977). However, Bellinger (1981) carried out a similar series of studies with generally negative results.

2.4. Signals controlling water and salt intake and metabolism

Ever since the work of Verney (1947) and Andersson (1953), osmolality of the extracellular fluid has been thought to be the chief signal in the control of dehydration thirst and renal diuresis. However, "osmotic" is a bad name for this control mechanism, as extracellular hyperosmolality does not itself activate thirst: it is the resultant cellular dehydration that is the signal. Both Verney, studying the control of pituitary release of vasopressin, the antidiuretic hormone (ADH), and Andersson, studying the control of water intake, concluded that these receptors were in the brain. However, subsequent work has also implicated the liver. The other major signal of water and sodium regulation, blood pressure, seems not to involve the liver in any direct way. Histamine, a recently proposed dipsogen, also apparently does not work through the liver (Kraly and Miller, 1982).

Haberich (1968) was the first to provide direct evidence for an osmoreceptor function in the liver and to link it to the antidiuretic effect of posterior pituitary secretions. In this experiment he co-infused hypertonic saline and water through either the hepatic-portal vein or the jugular vein. The resultant mix left osmolality unchanged in the general circulation. Whether diuresis or anti-diuresis was seen related to whether the solution went via the portal vessel, not the jugular.

Attempts to replicate these results, primarily in dogs, produced successes (Lydtin, 1969) and failures (Schneider, Davis, Robbs, Baumber, Johnson and Wright, 1970; Glasby and Ramsay, 1974; Kapteina, Motz, Schwartz-Porsche and Raver, 1978). More recent work using a more direct measure of ADH levels (Chwalbinska-Moneta, 1979) has been successful. Moreover, work done on rabbit (Ishiki, Morita and Hosomi, 1991) and rat (Cobashi and Adachi, 1988) clearly supports the Haberich conjecture. Comparisons of hepatic-portal and systemic veins showed greater hepatic effects of hypertonic solutions on renal nerve activity (a modulator of vasopressin release in rabbits). These studies were better than the earlier ones at mitigating stress effects. Also the measurements in these newer studies were more reliable than the rate of urine production upon which the earlier studies depended, which is difficult to control.

The role of hepatic osmoreception in behaviour appears to be weak in the case of water intake but strong if one examines salt appetite and satiety. Total subdiaphragmatic vagotomy does reduce water intake overall and lower sensitivity to osmotic changes. The best evidence is that this may have involved the gastrointestinal tract, however, and the gastric branch of the vagus rather than the liver and the hepatic vagus. The evidence for this is that the effects on water intake of total vagotomy were matched most closely by severing the gastric branch. Hepatic vagotomy had only very small effects on regulation of water balance (Smith and Jerome, 1983). Tordoff, Hopfenbeck and Novin (1981) saw no effect of hepatic vagotomy on water intake or its response to dehydration. On the other hand, when Bertino and Tordoff (1988) induced a salt depletion and hence a salt-hungry rat, hepatic infusion of sodium chloride was very effective at satiating salt intake and this effect depended on an intact hepatic branch of the vagus. Earlier, Blake and Lin (1978) showed that the sodium concentration that rats preferred to drink was better related to what was infused portally than it was to the systemic infusion.

2.5. Brain nuclei and pathways for hepatic signals

Visceral afferents travel in the brain in close proximity to and overlapping with gustatory afferents and some may even converge on the same cells. Thus, when a previously unknown major relay nucleus for taste was located in the parabrachial nucleus (Norgren and Leonard, 1971), it inspired many people interested in visceral information to begin a search of the central nervous system in close proximity to the gustatory areas for the visceral afferent pathways. One reason was that one of the gustatory nerves was the Xth cranial nerve, the vagus, which is also a, or probably the, major visceral afferent nerve. The gustatory map proved a useful guide for tracking the visceral vagal projections beyond the first relay in the nucleus of the solitary tract (NTS).

Our first attempt to find visceral pathways in the forebrain was to look in the ventrobasal complex of the thalamus adjacent to the thalamic taste area (TTA). Using microelectrode techniques, we found many cells whose activity was clearly an orderly response to the hypertonic solutions given via the hepatic-portal vein, while inferior vena cava infusions were without effect. At the end of each experiment, we injected horseradish peroxidase which is picked up by axon endings and usually transported back to the cell body which becomes densely coloured when the peroxidase has a suitable substrate. In this way we could label clusters of cells connected to the area in which we were recording (Fig. 2.1). In the case of the ventrobasal complex (VBC), there was good staining of the parabrachial nucleus (PBN), evidence of a substantial connection between the two nuclei. The use of other electrophysiological techniques tells us that signals go from the PBN to the VBC. Other labelling experiments linked the PBN to the NTS, the afferent traffic going from the NTS to the PBN.

2.5.1. *Central hepatic projections of water and salt regulation*

The supra-optic nuclei (SON) are on the border between the pre-optic area and the hypothalamus and sit bilaterally on the dorsal and lateral area of the optic chiasm. They secrete vasopressin in the face of increased osmolality and dehydration. They receive this information directly from the NTS projections, although these go mostly to the parvocellular portion of SON which is not directly involved in vasopressin release. The NTS also projects to the reticular formation (A2) in the brain stem and this area projects monosynaptically to the magnocellular neuroendocrine cells of the SON. There are also connections between NTS and the hormone-secreting cells of the paraventricular nucleus (PVN). Although many cells of the NTS project monosynaptically to the SON cells, more have synaptic connections to the PBN which then projects to the SON monosynaptically.

The solid evidence for osmosensitivity of the liver, and the observation of similar responses specifically to Na^+, support a role for these hepatic receptors in the vasopressin's regulation of renal diuresis. The rather direct neural pathway from liver receptors adds weight to the argument. Finally, while still considered above criticism by many, Bie (1980) has pointed out many flaws in the original work on central osmoreceptors. This leaves open the possibility that other systems participate in the modulation of vasopressin release.

The most likely function of the hepatic system is to act as a "feed forward"

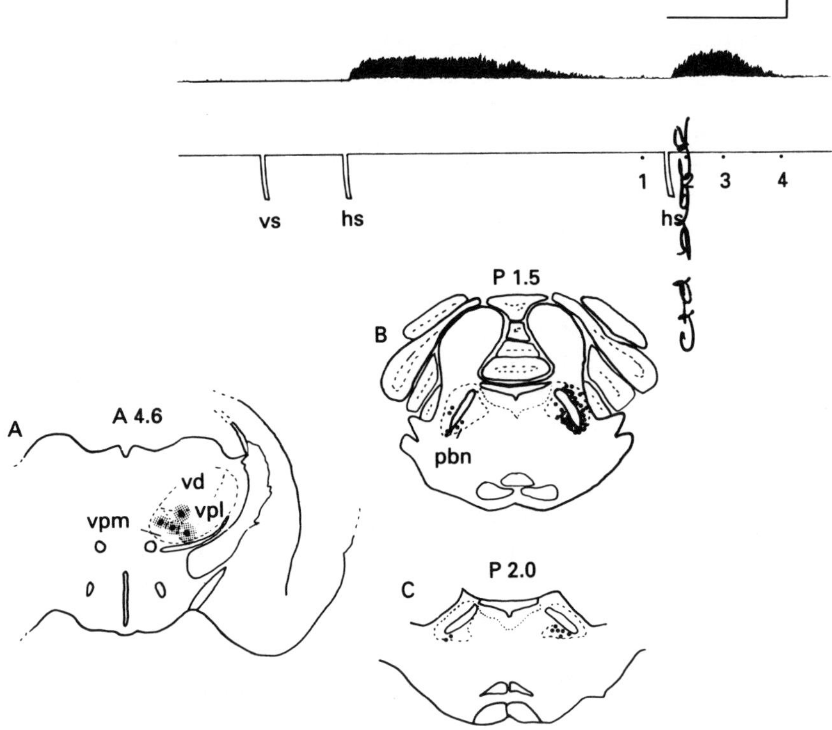

FIG. 2.1.   Single-cell response in the ventral basal complex of the thalamus to vena-caval or hepatic-portal infusion of salt solution. Insert A shows by the dark circles the locus of responsive cells. The stippled area surrounding these loci show the spread of injected HRP. B and C show the pontine and brain stem loci which project to the responsive area. These are respectively in B the PBN and in C the dorsal vagal complex. (Rogers *et al.*, 1979; reproduced with permission of the *Journal Of the Autonomic Nervous System.*)

satiation mechanism, anticipating the restoration of body fluid compartments and inhibiting further drinking (Fig. 2.2). If the animal stopped only when total body fluids were restored to normal, it is likely that disruptively positive water balance would result (Rogers and Novin, 1983).

2.5.2. *Central pathways for hepatic control of food intake*

The first relay nucleus for taste is at the rostral pole of the NTS and visceral afferents are in the caudal portion. In our hands, no cell in the NTS responded to both types of input. Reports to the contrary by others are probably based on recording in the subjacent reticular formation (Bereiter, Berthoud and Jeanrenaud, 1981). There are, however, cells in the NTS that show convergence between hepatic afferents and central gluco-receptors (Adachi, 1984). We also

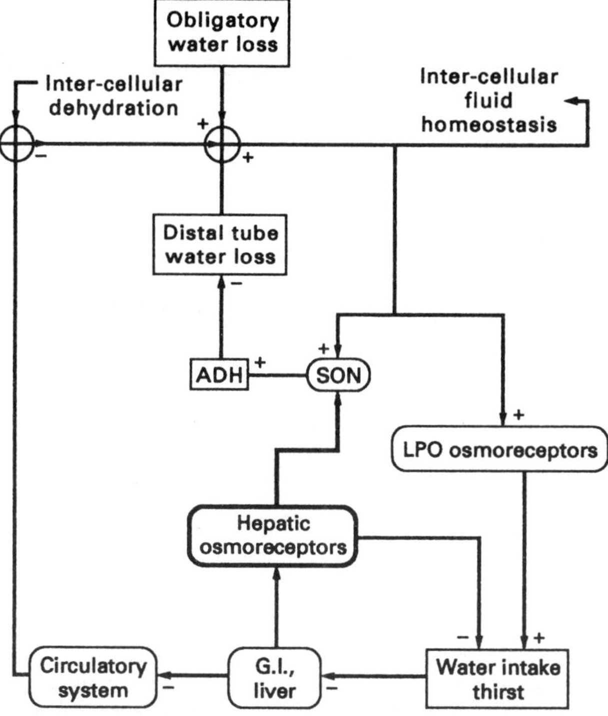

FIG. 2.2. A schematic of the role that a hepatic osmoreceptors could play on the regulation of the water intake and metabolism. By the liver's position between the gastro intestinal tract and the systemic circulation, these receptors prevent the overconsumption of water and the consequent dilution of body fluids and overexpansion of the intracellular compartment. Having an hepatic osmoreceptor allows the animal to stop drinking before the actual intracellular volume deficit is repaired. The hepatic osmoreceptors stop the drinking when their monitors report that the restoration will be accurate if the animal stops drinking now. This may well be before there has been any large increase in cell water.

see convergence between central glucoreceptors and hepatic glucosensitive afferents at the hypothalamic level (Fig. 2.3).

The PBN receives a large afferent input from the NTS. These PBN cells can be divided into two types, those that receive one modality but not both and those cells on which visceral and gustatory inputs converge. The non-convergent cells can be considered part of the classic sensory pathways, more or less preserving the purity of each modality. The gustatory afferents would continue to the TTA and cortical taste area.

There is now evidence that the PBN cells bifurcate, with one branch forming the connections to the TTA. Another branch carries fibres to the hypothalamus and limbic system. The most prominent of these pathways is that to the PVN. One can think of the brainstem and pontine areas as the locus for control of the

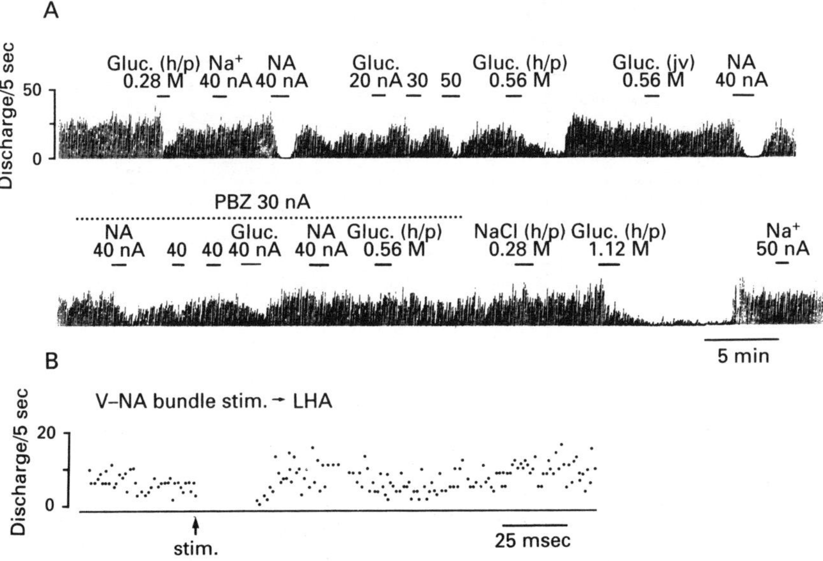

FIG. 2.3.   Rate-meter recording of a single unit in the lateral hypothalamic area identified as a glucose-sensitive unit by iontophoresis. This demonstrates that a single cell receives information from both liver and central nervous system gluco-receptors. It also demonstrates that the pathway from the liver to the hypothalamus is mediated by some adrenergic transmitter, as phenoxybenzamine (PBZ) reduced the effectiveness of hepatic portal glucose. Moreover, stimulation of the ventral noradrenergic bundle inhibited unit activity as did iontophoreses. (From Shimizu *et al.* (1983); reproduced by permission of *Brain Research*.)

Abbreviations: NA, noradrenaline; Na + , sodium; nA, nanoamps.

most basic feeding system, with oral stimulation modulated by visceral control (Fig. 2.4). In the decerebrate rat, the pons is separated from the forebrain. Norgren, Grill and Pfaffmann (1977) showed that these animals cannot feed spontaneously but they do show distinctive head and facial movements that are essentially the same as those shown by intact rats when something just ingested is accepted or rejected. The decerebrate rat shows through these movements what might be considered to be the two basic reflexes of feeding in the whole animal. One reflex is head and mouth movements of acceptance in response to a sugar solution, complemented by a very different but equally characteristic movement of rejection when stimulated by the taste of quinine. The second basic form of ingestive behaviour is displayed when the decerebrate rat that has been given a lot of sugar begins to display rejection movements instead of ingestive patterns; in other words, it shows the process of satiation. The neural circuitry at the pontine level provides the ability to detect and discriminate tastes and also the ability to combine visceral and gustatory inputs.

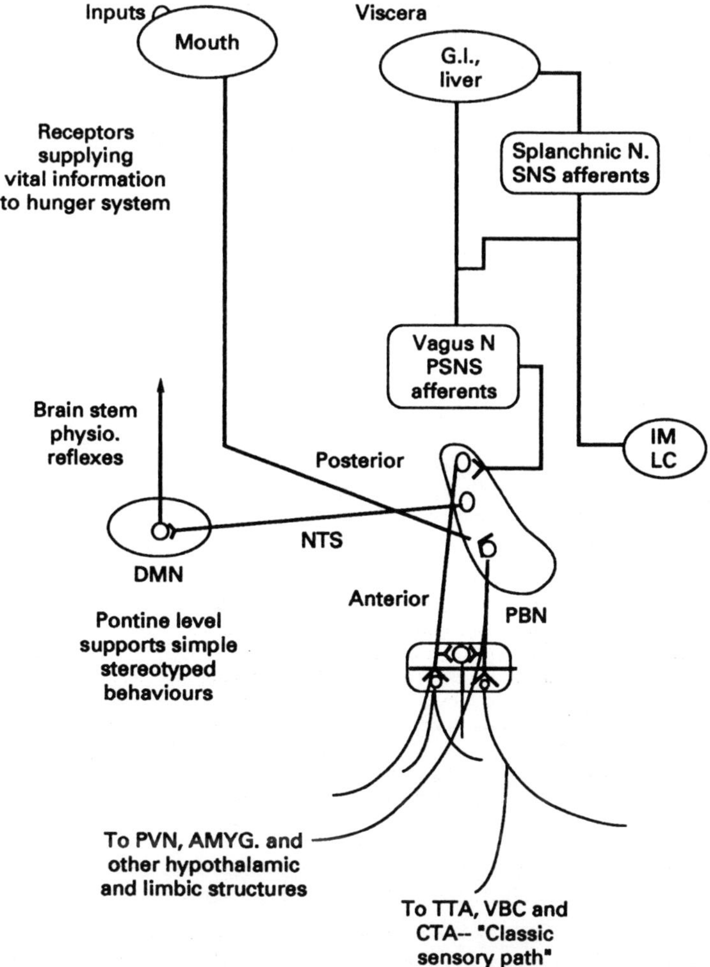

FIG. 2.4.   A simplified view of the neural substrates for a basic or primitive feeding system (largely from Norgren, Grill and Pfaffman, 1977). After decerebration, the animals were still able to show elements of feeding behaviour similar to the intact rat. At the pontine level and below, the neural apparatus exists for discriminating good from bad tastes. In addition the animal can integrate that information with information about the nutritional status of visceral organs.

Abbreviations: AMYG, amagdala; CTA, cortical taste area; DMN, dorsal motor nucleus of the vagus; IMLC, intero-medial lateral column; NTS, nucleus tractus solitarius; PBN, para-brachial nucleus; TTA, thalamic taste area; VBC, ventro-basal complex.

2.6. Conclusions

The liver is in the position to make an early determination of nutrient and non-nutrient substances entering the tissues of the body. It can certainly warn of overloads (and can only deal with some of them itself). The relatively few synapses between liver receptors and the hypothalamic endocrine cells of PVN and SON support the expectation that liver receptors will prove to have an important role in the regulatory control of hunger, possibly thirst and certainly metabolism (Ricardo and Koh, 1978). The more data are sought, the clearer will our view become of the information garnered from hepatic receptors.

References

Adachi, A. (1981). Electrophysiological study of the hepatic vagal protection to the medulla. *Neuroscience Letters*, **7**, 19–23.

Adachi, A. (1984). Convergence of hepatoportal glucose-sensitive afferent signals to glucose-sensitive units within the nucleus of the solitary tract. *Neuroscience Letters*, **46**, 215–218.

Andersson, B. (1953). The effect of injections of hypertonic NaCl solutions in different parts of the hypothalamus of goats. *Acta Physiologica Scandinavica*, **28**, 188–201.

Bellinger, L. L. (1981). Commentary on "The current status of hepatostatic theory of food intake control". *Appetite*, **2**, 144–145.

Bereiter, D. A., Berthoud, H. R. and Jeanrenaud, B. (1981). Chorda tympani and vagus nerve convergence onto caudal brainstem neurons in the rat. *Brain Research Bulletin*, **7**, 261–266.

Bertino, M. and Tordoff, M. G. (1988). Sodium depletion increased rats' preferences for salted food. *Behavioral Neuroscience*, **102**, 565–573.

Bie, P. (1980). Osmoreceptors, vasopressin, and control of renal water excretion. *Physiological Reviews*, **60**, 961–1048.

Blake, W. D. and Lin, K. K. (1978). Hepatic portal vein infusion of glucose and sodium solution and the control of drinking in the rat. *Journal of Physiology*, **274**, 129–139.

Booth, D. A. (1972). Postabsorptively induced suppression of appetite and energostatic control of feeding. *Physiology and Behavior*, **9**, 199–202.

Booth, D. A., Toates, F. M. and Platt, S. V. (1976). Control system for hunger and its implications in animals and man. In D. Novin, W. Wyrwicka and G. A. Bray (Eds.) *Hunger* (pp. 127–143). New York: Raven Press.

Cahill, G. (1977). Obesity and diabetes. In G. Bray (Ed.), *Recent Advances in Obesity Research*, Vol. II (pp. 101–110). London: Newman Publishing.

Campfield, L. A., Brandon, P. and Smith, P. J. (1985). On-line continuous measurement of blood glucose and meal pattern in free-feeding rats: the role of glucose in meal initiation. *Brain Research Bulletin*, **14**, 605–616.

Cannon, W. B. and Washburn, A. L. (1912). An explanation of hunger. *American Journal of Physiology*, **29**, 441–454.

Chwalbinska-Moneta, J. (1979). Role of hepatic portal osmoreception in the control of ADH release. *American Journal of Physiology*, **236**, E603–E609.

Deutsch, J. A., Young, W. G. and Kalogeris, T. J. (1978). The stomach signals satiety. *Science*, **201**, 165–166.

Fitz, J. O. and Scharschmidt, B. F. (1987). Regulation of transmembrane electrical potential gradient in rat hepatocytes *in situ*. *American Journal of Physiology*, **252**, G56–G64.

Friedman, M. (1991). Metabolic control of calorie intake. In. M. I. Friedman, M. G. Tordoff and M. R. Kare (Eds.), *Chemical Senses*, Volume 4: *Appetite and Nutrition* (pp. 19–38). New York: Marcel Dekker.

Glasby, M. A. (1974). Hepatic osmo-receptors? *Journal of Physiology*, **243**, 765–776.

Haberich, F. J. (1968). Osmoreception in the portal circulation, *Federation Proceedings*, **27**, 1137–1141.

Hermann, G., Kohlermann, N. and Rogers, R. (1983). Hepatic-vagal and gustatory afferent interactions in the brainstem of the rat. *Journal of the Autonomic Nervous System*, **9**, 477–495.

Ishika, K., Morita, H. and Hosomi, H. (1991). Reflex control of renal nerve activity originating from the osmoreceptors in the hepato-portal region. *Journal of the Autonomic Nervous System*, **36,** 139–148.

Kapteina, F. W., Motz, W., Schwartz-Porsche, D. and Raver, O. H. (1978). Comparison of renal responses to 5% saline infusions into ven portae and vena cava in conscious dogs. *Pflugers Archiv/European Journal of Physiology*, **374,** 23–29.

Kraly, F. S. and Miller, L. A. (1982). Histamine-elicited drinking is dependent upon gastric vagal afferents and peripheral angiotensin II in the rat. *Physiology and Behavior*, **28,** 841–846.

Kobashi, M. and Adachi, A. (1988). A direct hepatic osmoreceptive afferent projection from nucleus tractus solitaritus to dorsal hypothalmus. *Brain Research Bulletin*, **20,** 487–92.

Louis-Sylvestre, J. and Le Magnen, J. (1980). Food deprivation induced parallel changes in blood glucose, plasma free fatty acids and feeding during two parts of the diurnal cycle in rats. *Neuroscience and Biobehavioral Reviews*, **4,** 17–23.

Lydtin, H. (1969). Untersuchungen uber mechanismen der osmo- und volumeregulation. II. Untersuchungen uber den einfluss intravenos, intraportal und oral zuge-fuhrter hypotoner kochsalzlosung und die diurese des hundes. *Z. Gesamte Exp. Med.*, **149,** 193–210.

Niijima, A. (1969). Afferent impulse discharges from glucoreceptors in the liver of the guinea pig. *New York Academy of Sciences*, **157,** 690–700.

Niijima, A. (1983). Glucose-sensitive afferent nerve fibers in the liver and their role in food intake and blood glucose regulation. *Journal of Autonomic Nervous System*, **9,** 207–210.

Novin, D. (1988). Gustatory and visceral modulation of feeding. In J. Morley, B. Sterman and J. Walsh (Eds.), *Nutritional Modulation of Neural Function* (pp. 67–86). New York: Academic Press.

Novin, D., O'Farrell, L., Acevedo-Cruz, A. and Geiselman, P. J. (1991). The metabolic bases for "paradoxical" and normal feeding. *Brain Research Bulletin*, **27,** 435–438.

Novin, D. and VanderWeele, D. A. (1979) Visceral involvement of feeding. In J. Sprague and A. Epstein (Eds.), *Progress in Psychobiology and Physiological Psychology* (pp. 193–241). New York: Academic Press.

Novin, D., VanderWeele, D. A. and Rezek, M. (1973). Infusion of 2-deoxy-D-glucose into the hepatic-portal system causes eating: evidence for peripheral glucoreceptors. *Science*, **181,** 858–860.

Norgren, R. and Leonard, C. M. (1971). Taste pathways in rat brainstem. *Science*, **173,** 1136–1139.

Norgren, R., Grill, J. and Pfaffmann, C. (1977). CNS projections of taste to the dorsal pons and limbic system with correlated studies of behavior. In Y. Katsuki, M. Sato, S. Takagi and Y. Oomura (Eds.), *Food Intake and Chemical Senses* (pp. 233–243). Japan: University of Tokyo Press.

Prechtl, J. and Powley, T. (1985). Organization and distribution of the rat subdiaphragmatic vagus and associated paraganglia. *Journal of Comparative Neurology*, **235,** 182–195.

Ricardo, J. A. and Koh, E. T. (1978). Anatomical evidence of direct projections from the nucleus of the solitary tract to the hypothalamus, amygdala and other forebrain structures in the rat. *Brain Research*, **153,** 1–26.

Rogers, C. and Novin, D. (1983). The neurological aspects of hepatic osmoregulation. In M. Epstein (Ed.), *The Kidney in Liver Disease* (pp. 337–350). New York: Elsevier Biomedical.

Russek, M. (1963). An hypothesis on the participation of hepatic glucoreceptors in the control of food intake. *Nature, London*, **197,** 79–80.

Scharrer, E. and Langhans, W. (1988). Metabolic and hormonal factors controlling food intake. *International Journal of Vitamin and Nutrition Research*, **58,** 249–261.

Schneider, E. G., Davis, J. O., Robbs, C. A., Baumber, J. S. and Wright, F. S. (1970). Lack of evidence for an hepatic osmoreceptor in conscious dogs. *American Journal of Physiology*, **218,** 42–45.

Shimazu, T. (1967). Glycogen synthetase activity in liver: regulation by the autonomic nerves. *Science*, **156,** 1256–1257.

Shimizu, N., Oomura, Y., Novin, D., Grijalva, C. and Coopers, P. (1983). Functional correlations between lateral hypothalamic glucose sensitive neurons and hepatic portal glucose sensitive units in the rat. *Brain Research*, **265,** 49–54.

Smith, G. P. and Jerome, C. (1983). Effects of total and selective abdominal vagotomies on water intake in rats. *Journal of Autonomic Nervous System*, **9**, 259–271.

Stricker, E. M., Rowland, M., Saller, C. F. and Friedman, M. I. (1977). Homeostasis during hypoglycemia: central control of adrenal secretion and peripheral control of feeding. *Science*, **196**, 79–81.

Tordoff, M., Hopfenbeck, J. and Novin, D. (1981). Hepatic vagotomy (partial hepatic denervation) does not alter ingestive responses to metabolic challenges. *Physiology & Behavior*, **28**, 417–424.

Verney, E. B. (1947). The antidiuretic hormone and the factors which determine its release. *Proceedings of the Royal Society of London, B.*, **135**, 25–106.

3

Hepatic afferents affecting ingestive behaviour

WENDY R. EWART

Gastrointestinal Science Research Unit, The London Hospital Medical College, London, England E1 2HA

3.1. Introduction

BY virtue of the liver's pivotal role in balancing the ebb and flow of essential nutrients, it has long been held that this organ occupies a central role in homeostasis (Novin, Chapter 2). Through mechanisms for detoxification, clearance, hormone inactivation and storage of metabolic intermediaries, the liver is able to adjust the composition of the blood in response to the changing demands of the internal environment. It performs this function with great effectiveness in relation both to short–term requirements (adaptation to fed and fasting states) and also in response to long-term changes in environment (such as periods of rapid growth or of impoverished dietary input).

In view of the ability of the liver to respond as a metabolic sensor, it has been supposed that it was also a sensate structure. Solid evidence for this supposition has been lacking, though, until the recent past. Efforts to prove the sensibility of hepatic structures have been hampered by the inadequacy of classical histological techniques to demonstrate the fine unmyelinated nerve fibres which characteristically innervate the liver. Since the advent of sophisticated anterograde and retrograde tracing techniques which can clearly demonstrate the afferent and efferent connections of visceral structures, the innervation of hepatic structures is now well established anatomically.

However, much remains to be understood about the identity, morphology and behaviour of the hepatic receptors themselves. Impressive progress has been made in this area by the use of electrophysiological techniques to record from single primary vagal afferent fibres. By such a method, it is possible to characterize different hepatic receptors; amongst those investigated in this way have been glucoreceptors, osmoreceptors and pressure/volume receptors.

Single unit recordings, made through microelectrodes stereotactically positioned in the brainstem, have also been utilized to learn about the fate of sensory information from the liver once it enters the central nervous system (CNS). It is not yet clear what roles this centrally directed information has. Influencing the control of food intake would have to be a likely possibility, as would other important integrative functions such as the control of blood volume and the neural control of both the endocrine and exocrine pancreas. There is evidence for the involvement of sensory information from the viscera, and from the liver in particular, in the elaboration of these complex homeostatic mechanisms at the level of the hypothalamus and beyond. In understanding the neural mechanisms responsible for ingestive appetites such as hunger and thirst, behavioural studies are of course indispensible to our appreciation of the role of hepatic afferents.

3.2. Neuroanatomy

3.2.1. Peripheral neuroanatomy

The liver is innervated by certain parts of the vagus, splanchnic and phrenic nerves. There are considerable species differences in the relative contributions of these nerves, particularly with respect to the contribution of the phrenic nerve. There appear to be two major points of entry to the liver for these nerves, associated with the hepatic artery and portal vein; there is abundant evidence for nerve terminals associated with these vascular structures (Reilly *et al.*, 1957). Although the majority of intrinsic hepatic nerves appear to associated with the vascular and biliary systems, there is accumulating evidence that hepatic parenchymal and other intra-lobular structures are also innervated (Skaaring and Bierring, 1976, 1977). It has been important to establish whether the intralobular areas of the liver are innervated because this has a bearing on whether hepatic receptors monitor the vascular supply to the liver or sense the contents of hepatocytes, giving a more integrated version of metabolic events.

The evidence for afferent nerve endings in the liver is sparse. It rests principally on structural similarities between nerve endings in the liver and known sensory receptors in other areas such as the carotid body, as well as on degeneration studies following nerve section below the level of sensory ganglia (for review see Sawchenko and Friedman, 1979). Three types of nerve endings have been described: bare nerve endings, within lobules and around central veins and bile ducts; Pacinian corpuscles (connective tissue septa); and splayed glomerular endings (reminiscent of those seen in the carotid body), found in association with the bile ducts (Tsai, 1958). The presumption that some or all of these structures serve as receptors rests, however, on evidence supplied by recording electrical activity in primary afferents from the liver.

3.2.2. Central connections

Anatomical studies in recent years have focused on the route which afferent information from the liver takes between the viscera and the CNS in the vagal nerves. The reasons for concentrating research effort on parasympathetic rather than sympathetic innervation are that the majority of the fibres in the vagus are afferent (75–90%) and, as with the stomach, the principal route for signalling stimuli such as movement and chemical composition is the vagus (for review see Grundy, 1988). The main role of the splanchnic innervation may be the relay of nociceptive information; transmission of specific messages from other visceral sensory receptors seems to be of lesser importance in this sympathetic autonomic route via the spinal cord to the brain.

The subdiaphragmatic trunks of the vagus nerve and their branches (Fig. 3.1) project to the primary vagal afferent relay area in the medulla, the nucleus of the solitary tract (NTS). Detailed information about the hepatic branch of the vagus nerve has been obtained using the axoplasmic tracer horseradish peroxidase (Rogers and Hermann, 1983; Norgren and Smith, 1988). With this method of labelling, the central distribution of the hepatic vagus has been described in elegant detail. The hepatic branch was reported to be the most difficult part of the vagus nerve to label successfully. There was only a scant labelling of terminals in the anterograde direction, no doubt reflecting the paucity of hepatic fibres in comparison with other visceral areas. Anterograde labelling in the NTS was almost exclusively distributed in the nucleus on the left side of the medulla. The greatest concentration of terminals lay around the midline, in the area of NTS subjacent to the area postrema (Fig. 3.2). Very little reaction product (that is, very few afferent terminals) was observed in the area postrema itself or in the medial NTS just dorsal to the dorsal motor nucleus of the vagus (DMNV).

Afferent fibres from different parts of the abdominal viscera overlap to a considerable degree within the brain stem; there are only small areas of separation between the input from different vagal branches. This suggests a possibility of functional convergence for the neural information being carried by these nerves, an hypothesis tested and confirmed by electrophysiological studies of neurones in this area of the brainstem (Appia *et al.* 1986). In contrast, there seems little opportunity for integration of this hepatic and gastric afferent information with that from gustatory information, since the taste projections are principally to more rostral parts of NTS (Hermann *et al.*, 1983).

As might be expected for afferent fibres from the alimentary viscera, the prime destination for ascending fibres from their relay in the caudal NTS is the hypothalamus (Ricardo and Koh, 1978), specifically the dorsomedial hypothalamus (DMH) and the paraventricular nucleus (PVN). Sensory information from the liver passes to the hypothalamus directly and via the parabrachial nucleus (PBN); this allows further opportunities for integration of the afferent profile from hepatic receptors with afferent information from enteric receptors. From the PBN, direct connections have been established with specific parts of

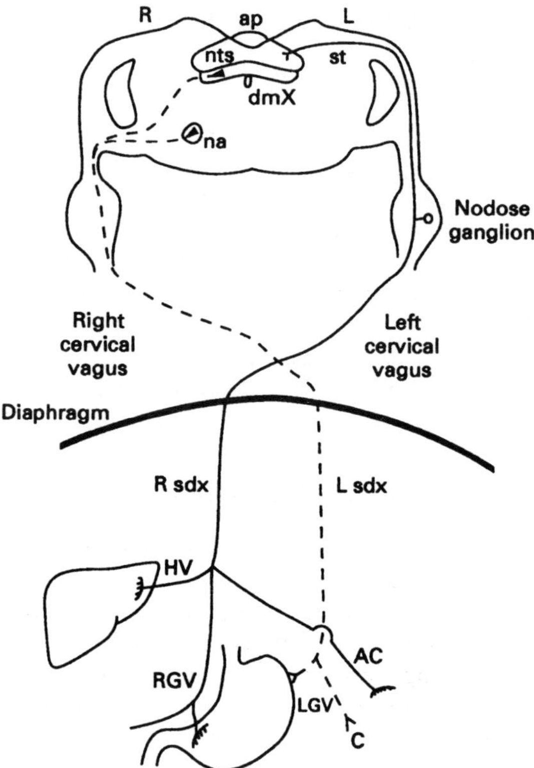

FIG. 3.1.   Distribution and central projections of the subdiaphragmatic branches of the vagus. Left (L) and right (R) subdiaphragmatic branches of the vagus, Xth cranial nerve (sdx); AC, accessory coeliac; C, coeliac; HV, hepatic vagus; LGV, left gastric vagus; RGC, right gastric vagus; ap, area postrema; dmX, dorsal motor nucleus; na, nucleus ambiguus; nts, nucleus of the solitary tract; st, solitary tract.

the hypothalamic area including ventromedial hypothalamus (VMH), DMH, PVN and the lateral hypothalamic area (LHA) (Norgren, 1978).

Beyond the hypothalamus, onward routes for hepatic information within the brain await further clarification. There is no doubt, however, that the destinations of this information include the association areas of cerebral cortex. This provides ample opportunity for sensory influences from the liver to enter into learned control of ingestion (Rolls, Chapter 9). Nevertheless, the mechanisms through which afferent information from hepatic receptors may be transduced into sensations referred to the abdomen or into appropriate forms of ingestive behaviour remain to be elucidated.

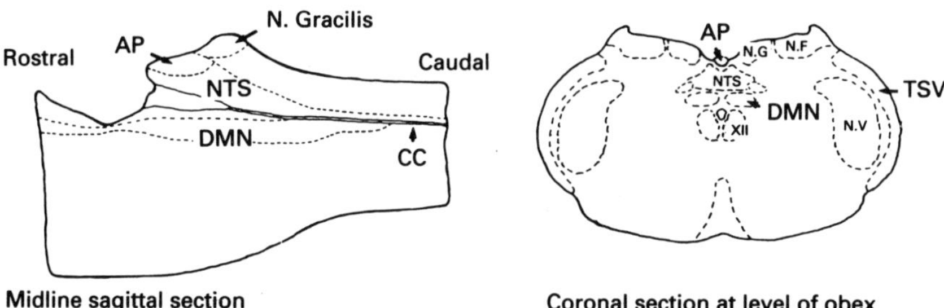

FIG. 3.2.   Sagittal and coronal sections of the rat brain stem to show the extent and position of brainstem structures involved in processing afferent information from the gastrointestinal tract. AP, area postrema; CC, corpus postrema; DMN, dorsal motor nucleus; NF, cuneate nucleus; NG, gracile nucleus; NTS, nucleus of the solitary tract; NV, trigeminal nucleus; TSV, trigeminal tract; XII, hypoglossal nucleus.

3.3. Neurophysiology of hepatic receptors

The most direct way of studying characteristics of hepatic receptor activation has been the electrical recording of unitary impulse activity in primary afferent fibres of the hepatic branch of the vagus nerve, almost a "classical" preparation. This technique has been used to great effect in determining the properties of enteric receptors, including those in the stomach and duodenum. Nerve filaments are isolated from the hepatic branch of the vagus under a dissecting microscope and the distal cut ends are placed on a pair of silver wire electrodes and immersed in a mixture of liquid paraffin and vaseline. Conventional signal-analysis techniques are used to gate out and display the firing in a single fibre. Stimuli are then applied to the region of the liver, usually by perfusion through the hepatic-portal vein using a non-occlusive cannula. Thus it is possible to mimic signals presented to the liver by absorption from the upper intestine. Their effects on hepatic receptor activity are assessed by observing changes in rate of the unitary firing of the hepatic vagal afferent fibre.

A wide range of stimuli has been investigated in this way and four main groups of hepatic receptors identified, osmoreceptors, ionic receptors, pressure/volume receptors and metabolic receptors, which include those responsive to glucose, amino acids and sugars (Table 3.1). It has also been postulated that there are thermoreceptors within the liver but these are less well documented. There has been some division of opinion as to the specificity of sensory modality of these receptors, but there is general agreement that a surprisingly wide range of factors are monitored by the liver. Undoubtedly this sensory information is of considerable importance in interpreting the chemical and mechanical environment of the abdominal viscera.

Most authors referred to in the table ascribe the changes in firing rate they observed to effects of the agonists on "receptors" *per se*, rather than being through indirect receptor activation. However, it may well be that insulin and

TABLE 1. *Evidence for hepatic receptors from primary vagal afferent recording*

Hepatic receptor	Sensitive to:	Reference
Osmoreceptive	Hyperosmolar NaCl	Niijima (1969)
Ionic	Na^+	Andrews and Orbach (1974)
	Na^+	Adachi *et al.* (1976)
Mechanical	Stretch	
(Pressure/volume)	[splanchnic]	Niijima (1977)
Metabolic	Glucose	Niijima (1969, 1983, 1984)
	Amino acid (arginine)	Tanaka *et al.* (1986)
	Insulin	Niijima (1988)
	Glucagon	Niijima (1988)
	Cholecystokinin	Niijima (1988)
	Endogenous sugar acids	Niijima (1988b)
	2-Deoxyglucose	Niijima (1984)
Thermoreceptive	Direct liver warming	Adachi (1974)

glucagon exert their effects on hepatic afferents indirectly, through their action on glucose metabolism. In addition, the effect of glucose in these experiments might be intraneuronal rather than on the nerve-ending membrane and so better described as acting on "glucose-sensitive neurones" than called 'glucoreceptor' action. Nevertheless, questions about the specificity and mechanism of these responses cast no doubt on the evidence that changes in the chemical composition of the hepatic portal vein are detected with great sensitivity by hepatic receptors and that this information is coded in changes in the pattern and frequency of action potentials carried to the CNS by vagal afferents directly from the liver.

Glucose-sensitive afferents have a consistent response, showing an inverse relationship between glucose concentration and firing rate (Niijima, 1983; Fig. 3.3). In addition, certain endogenous sugar acids, shown to suppress food intake (Oomura, 1987), also decrease the activity of glucose-sensitive hepatic afferents (Niijima, 1988a). 2-Deoxy-D-glucose (an inhibitor of glucose metabolism) causes vagally dependent eating (Novin *et al.*, 1973), and increases activity in hepatic vagal afferents. These pieces of evidence taken together suggest that sensory information from hepatic afferents is indeed involved in the metabolic control of food intake (see Novin, Chapter 2). It should be noted, however, that food intake stimulated by blockade of glucose metabolism in rats has been reported to be less dependent on vagal projections to the brainstem than feeding elicited by blockade of fatty-acid oxidation (Rutter and Taylor, 1990) because 2DG also acts in the brainstem directly.

3.4. Central processing of information from hepatic afferents

Primary hepatic vagal afferent fibres terminate in the NTS in the brain stem; this is the first point in the onward transmission of sensory signals from the abdom-

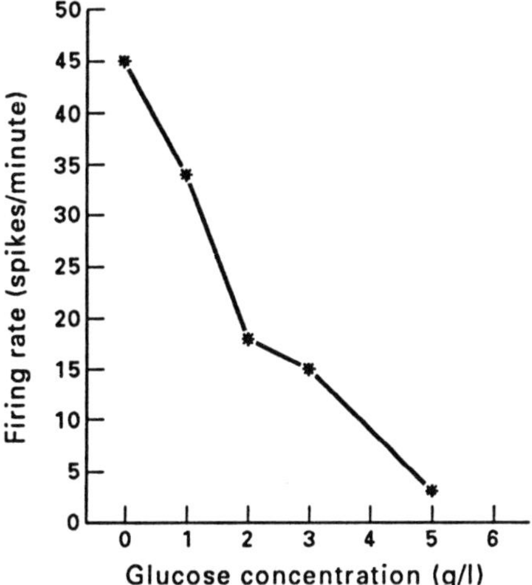

FIG. 3.3.   Inverse relationship between firing rate of a primary hepatic afferent vagal fibre and perfused into the hepatic portal vein (drawn from data presented in Niijima, 1983).

inal viscera at which synaptic interaction with other pathways becomes possible. Experimentally, it is an area which has aroused considerable interest as this is the stage at which neurotransmitters and neuromodulators might have a role to play in refining the profile of visceral afferent information within the CNS.

Experimental techniques to explore brainstem handling of afferent information have included electrophysiology, sophisticated morphological studies and various types of lesions and pharmacological interventions combined with behavioural or intake measurements. Neurophysiological experiments have concentrated on the responses of single neurones in the brainstem to both physiological and electrical activation of afferent pathways. Investigators in this field commonly find that it is difficult to retain stable recording conditions long enough to characterize the behaviour of each neurone fully. This is primarily because the brainstem, being an extra-cranial structure, is especially difficult to isolate from movements such as respiration. As a result, the success rate of unitary recording in this type of study is particularly low, which might explain why there are so few reports of brainstem processing of visceral afferent information.

Amongst the afferent projections investigated in the brainstem has been that from the activation of osmoreceptors. Kahrilas and Rogers (1985) described in detail the responses of individual neurones in the left medial NTS to portal infusion of water or hypertonic saline. The predominant response was a persistent change in firing rate lasting some 20 minutes or more described as "on–off" responses. These neurones were excited by portal perfusion of hypertonic saline and inhibited by perfusion of water. Interestingly, the direction of response, "on" or "off", could be switched rapidly by alternating perfusates. These responses were sensitive to a change in portal sodium ion concentration of less than 1% and were sodium specific. It was suggested that these neurones act as thirst and diuretic threshold detectors.

Remarkably similar responses patterns were seen by Appia *et al.* (1986) who demonstrated integration at the brainstem level of sensory information from hepatic and gastric receptors (Fig. 3.4). Glucose-sensitive afferents were activated by portal infusion and units responding to this stimulus were also tested with gastric distension by the inflation of a small water-filled balloon placed in the stomach via a gastrostomy; this technique has previously been shown to activate vagal gastric mechanoreceptors. Figure 3.5 shows an example of a unit which responded to both gastric distension and hepatic glucose perfusion, thereby illustrating convergence of sensory information from different parts of the viscera at this first relay in the CNS. Like the responses to hepatic stimulation described by Kahrilas and Rogers (1985), units in these studies displayed the unusual characteristic of being "switched" on or off by different perfusates. In this example, the neurone was activated within a few seconds of the gastric distension response being applied and this activation continued for many minutes. Yet it was possible to "switch off" this activity by portal infusion of isotonic glucose, while isotonic saline did not have any effect.

Observations were also made of units which appeared to respond to the volume of perfusate. Figure 3.6 shows an example of one of these units which showed an impressive and prolonged change in firing rate following a slow injection (given over 1 minute) of a small volume of isotonic saline (0.05 or 0.1 ml). The role of hepatic pressure/volume sensitive receptors is not clear, but the observation that their activation is monitored so accurately in the brainstem, indicates that the fluid dynamics of absorption from the GI tract provides an important neural signal. Information from hepatic pressure/volume receptors might contribute to the control of blood volume (Strandhoy and Williamson, 1970).

An alternative to physiological activation of hepatic vagal afferents is selective electrical stimulation. Although this technique loses much of the functional meaning of the response (by initiating a non-modality specific train of action potentials), it gains in other respects. For example, the antidromic activation of a brainstem neurone, along with testing for "collision" between spontaneous and stimulated impulses, can show that the brainstem unit being studied is an output neurone, that is, a vagal pre-ganglionic motor neurone. This enables the electrophysiological identification of second-order sensory relay neurones by exclusion; that is, if a unit fails to meet the criteria of antidromic activation and collision testing, a presumptive identification of a second (or higher) order sensory neurone can then be made.

Using these techniques, recordings have been made from medullary neurones which responded to portal perfusion of isotonic glucose, hypertonic saline and

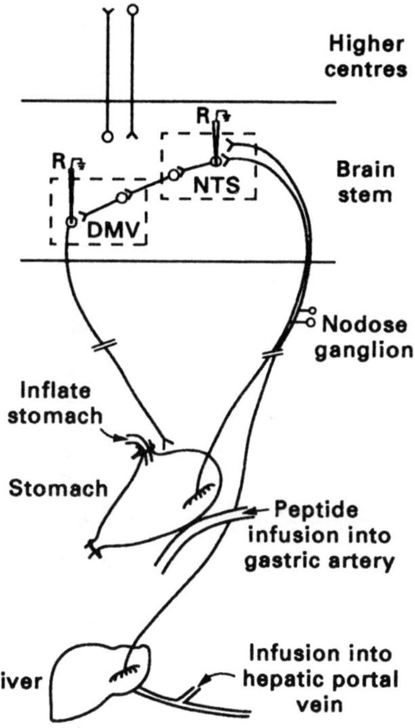

FIG. 3.4.   Experimental model used to study the responsiveness of brainstem neurones to perfusion of the hepatic portal vein with isotonic glucose and isotonic saline.

water in ways that were very similar to the responses previously observed in hepatic afferents (Adachi, 1981). More recently, a small number of units in NTS which received afferent input from the hepatic vagus and projected to PBN, were shown to influenced by electrical stimulation of the hepatic vagus, and some of these units were responsive to hepatic osmoreceptor activation (Kobashi and Adachi, 1986). Some neurones responding to portal stimulation were shown to be glucose-sensitive in themselves (Adachi *et al.* 1984), by topical application of glucose solution from the recording microelectrode, using iontophoresis. Glucose-sensitive neurones have also been found in the hypothalamus (Oomura and Yoshimatsu, 1984). Both sets of neurones are presumably part of the neural mechanisms contributing to glucose homeostasis.

The hypothalamus coordinates a number of homeostatic mechanisms through pituitary and brainstem outflows, not least of which is osmoregulation and the control of fluid intake. NTS neurones reactive to hepatoportal saline infusion project directly to the hypothalamus (Schmitt, 1973; Kobashi and Adachi, 1988). The representation of hepatic receptors at this level of the CNS further underlines the importance of monitoring the chemical composition and fluid dynamics of the portal blood supply.

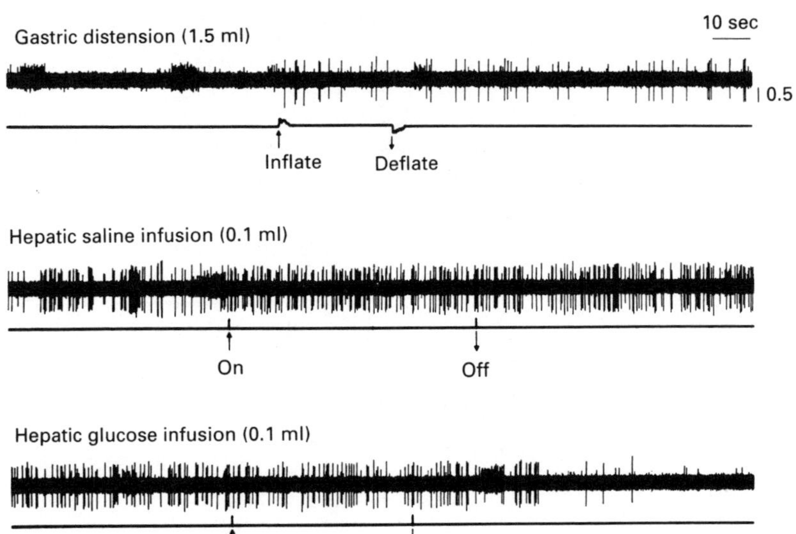

FIG. 3.5.   Example of a neurone excited by gastric distension, inhibited by hepatic portal-vein perfusion of isotonic glucose solution and unaffected by hepatic perfusion of isotonic saline, thereby demonstrating convergence of sensory information from different regions of the abdominal viscera at the level of the medulla. This figure shows three sections of recording separated in time. Horizontal lines below the spike traces indicate periods of stimulation. *Upper trace*: following inflation of the gastric balloon activity was evoked in a hitherto quiescent neurone. Another neurone is seen firing in small bursts; this neurone's activity continued through the period of recording but did not appear to be markedly affected by the stimuli. *Middle trace*: infusion of a small volume of saline had no appreciable affect on the activity of this neurone. *Bottom trace*: infusion of 0.1 ml of isotonic glucose solution through the hepatic portal vein inhibited firing of the neurone. The sequence of events was consistently repeatable.

Clearly then, there is accumulating evidence that modality-specific sensory information from the liver is projected centrally in sufficient detail and clarity to be able to make a significant contribution to the control of ingestion. The use of single-unit recording techniques combined with peripheral stimulation and central modulators is an experimental approach that is indispensable to further progress in understanding the role of afferent information in the control of ingestive behaviour. There is relatively little experimentation in this field of CNS neurophysiology and, as yet, no use of behavioural techniques that can operate on a relevant timebase under practicable recording conditions. There is ample scope, therefore, for further study in this aspect of the neurophysiology of ingestion.

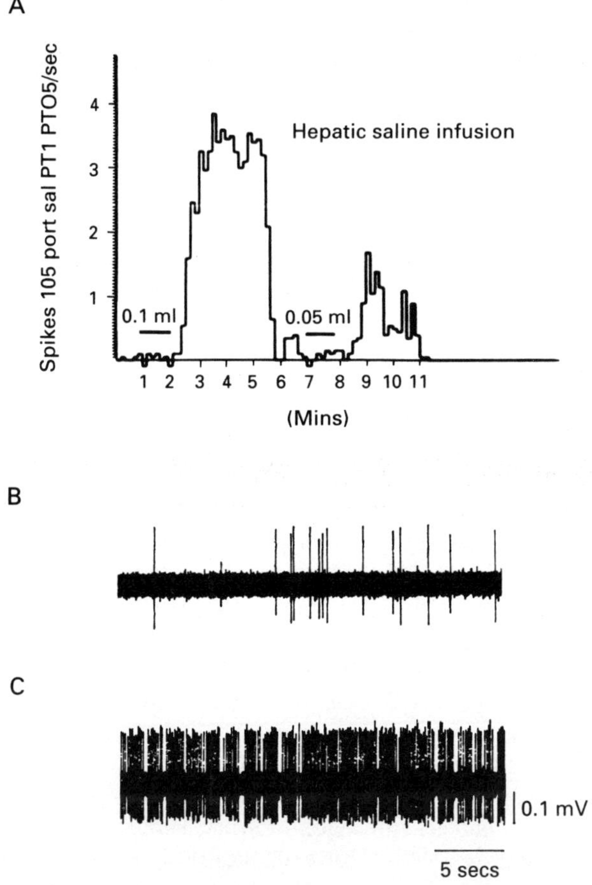

FIG. 3.6.   An example of a neurone recorded in the dorsal vagal nucleus that was inhibited by gastric distension and responded to infusion of small volumes of isotonic saline through the hepatic portal vein. *Panel A*: Computer-generated spike frequency histogram of the changes in firing rate during two consecutive infusions of saline. Horizontal bars indicate periods of infusion. The latency of the response was long (greater than 60 seconds after the onset of infusion). The magnitude of the change in firing rate was related to the volume injected. *Panel B*: Extracellularly recorded spike trace, showing the baseline firing rate prior to hepatic saline infusion. *Panel C*: Section of spike trace showing the increase in firing rate observed in response to hepatic infusion of isotonic saline.

3.5. Autonomic functions of sensory information from the liver

While we lack a clear understanding of the roles of hepatic afferents in the central neural pathways in the control of food intake, it would be wise to consider also the other functions of hepatic sensors, namely in autonomic control.

Lee (1985) has reported a vago-vagal reflex linking the excitation of hepatic vagal afferent neurones with the modulation of gastric vagal efferent neurones. In these studies, electrical stimulation of the cut central end of the hepatic vagus nerve resulted in the inhibition of spontaneous gastric contractions; on the other hand, in the absence of any spontaneous gastric motility, hepatic vagal stimulation elicited contractions. This work confirmed previous observations of the link between hepatic afferents and gastric motility (Sakaguchi and Miyaoka, 1981). Hepatic metabolism has a role in this reflex response. Granneman and Friedman (1980) showed that insulin-induced changes in gastric motility were mediated through the hepatic vagus nerve. Thus, perturbation of glucose homeostasis has widespread effects, feeding back to activity of the digestive tract.

Information from hepatic afferents may well also be involved in the control of the endocrine pancreas and perhaps its exocrine digestive secretions as well. Studies using the neurotoxin capsaicin, which selectively destroys fine unmyelinated nerve fibres, have elegantly demonstrated the need for vagal afferent information from glucose receptors in the hepatic portal vein in order to achieve insulin-induced increase in adrenaline secretion (Amann and Lembeck, 1986). However, control of insulin secretion is not the exclusive domain of hepatic vagal afferents: Berthoud *et al.* (1983) have shown that any of the subdiaphragmatic branches of the vagus nerve support neurally mediated insulin release.

The balance between insulin and its hormonal opponent, glucagon, has attracted the attention of workers whose primary interest has been in mechanisms involved in the control of food intake. Glucagon-induced suppression of food intake has been well documented and appears to be mediated primarily by an action on the liver (Weick and Ritter, 1986). More recent evidence pinpoints this effect to hepatic vagal afferents, since the destruction of vagal afferent terminals in the brainstem impairs glucagon-induced suppression of food intake (Weatherford and Ritter, 1988). However, these lesions spared the satiety effect induced by cholecystokinin, a response mediated by gastric vagal afferent neurones. Hence, sensory information from the liver is not simply travelling in an undifferentiated package with that from gastric afferents. The judicious combination of specific lesions and localized chemical manipulations with behavioural studies complements neurophysiological studies advancing our understanding of these intricate integrative mechanisms.

References

Adachi, A. (1974). Afferent neural unit of the hepatic branch of the vagus nerve. *Journal of the Physiological Society of Japan (Abstract)*, **36**, 310.

Adachi, A., Niijima, A. and Jacobs, H. L. (1976). An hepatic osmoreceptor mechanism in the rat: electrophysiological and behavioural studies. *American Journal of Physiology*, **231**, 1043–1049.

Adachi, A. (1981). Electrophysiological study of hepatic vagal projection to the medulla. *Neuroscience Letters*, **24**, 19–23.

Adachi, A., Shimuzu, N., Oomura, Y. and Kobashi, M. (1984). Convergence of hepatoportal glucose-sensitive units within the nucleus of the solitary tract. *Neuroscience Letters*, **46**, 215–218.

Amann, R. and Lembeck, F. (1986). Capsaicin sensitive afferent neurons mediate the insulin-induced increase in adrenaline secretion. *Naunyn-Schmiedeberg's Archives of Pharmacology*, **334**, 71–76.

Andrews, W. H. H. and Orbach, J. (1974). Sodium receptors activating some nerves of perfused rabbit livers. *American Journal of Physiology*, **227**, 1273–1275.

Appia, E., Ewart, W. R., Pittam, B. S. and Wingate, D. L. (1986). Convergence of sensory information from abdominal viscera in the rat brain stem. *American Journal of Physiology*, **251**, G169–175.

Bethoud, H. R., Niijima, A., Sauter, J. -F. and Jeanrenaud, B. (1983). Evidence for a role of the gastric, coeliac and hepatic branches in vagally stimulated insulin secretion in the rat. *Journal of the Autonomic Nervous System*, **7**, 97–110.

Granneman, J. and Friedman, M. I. (1980). Hepatic modulation of insulin-induced gastric acid secretion and EMG activity in rats. *American Journal of Physiology*, **238**, 346–352.

Grundy, D. (1988). Speculations on the structure/function relationships for vagal and splanchnic afferent endings supplying the gastrointestinal tract. *Journal of the Autonomic Nervous System*, **22**, 175–180.

Hermann, G. E., Kohlerman, N. J. and Roger, R. C. (1983). Hepatic-vagal and gustatory interactions in the brainstem of the rat. *Journal of the Autonomic Nervous System*, **9**, 477–495.

Kahrilas, P. J. and Rogers, R. C. (1984). Rat brainstem neurons responsive to changes in portal blood sodium concentration. *American Journal of Physiology*, **247**, 792–799.

Kobashi, M. and Adachi, A. (1986). Projection of nucleus solitarius units influenced by hepatoportal afferent signal to parabrachial nucleus. *Journal of the Autonomic Nervous System*, **16**, 153–158.

Kobashi, M. and Adachi, A. (1988). A direct hepatic osmoreceptive afferent projection from nucleus tractus solitarius to dorsal hypothalamus. *Brain Research Bulletin*, **20**, 487–492.

Lee, K. C. (1985). Reflex suppression and initiation of gastric contractions by electrical stimulation of the hepatic vagus nerve. *Neuroscience Letters*, **53**, 57–62.

Niijima, A. (1969). Afferent impulse discharges from osmoreceptors in the liver of the guinea-pig. *Science*, **166**, 1519–1520.

Niijima, A. (1977). Afferent discharges from venous pressoreceptors in liver. *American Journal of Physiology*, **232**, C76–C81.

Niijima, A. (1982). Glucose-sensitive afferent nerve fibres in the hepatic branch of the vagus nerve in the rat. *Journal of Physiology*, **322**, 315–323, 1982.

Niijima, A. (1983). Glucose-sensitive afferent fibres in the liver and their role in food intake and blood glucose regulation. *Journal of the Autonomic Nervous System*, **9**, 207–220.

Niijima, A. (1984). The effect of D-glucose on the firing rate of glucose-sensitive vagal afferents in the liver in comparison with the effect of 2-deoxy-2-D-glucose. *Journal of the Autonomic Nervous System*, **10**, 255–260.

Niijima, A. (1988a). The effect of gastro-entero-pancreatic hormones on the activity of vagal hepatic afferent fibres. In W. Hamann and A. Iggo (Eds.), *Progress in Brain Research*, Vol. 74 (pp. 155–160). Elsevier.

Niijima, A. (1988b). The effect of endogenous sugar acids on the afferent discharge rate of the hepatic branch of the vagus nerve in the rat. *Physiology and Behavior*, **44**, 661–664.

Norgren, R. (1978). Projections from the nucleus of the solitary tract in the rat. *Neuroscience* **3**, 207–218.

Norgren, R. and Smith, G. P. (1988). Central distribution of subdiaphragmatic vagal branches in the rat. *Journal of Comparative Neurology*, **273**, 207–223.

Novin, D., Vander Wheele, D. A. and Russek, M. (1973). Infusion of 2-deoxy-2-D-glucose into the hepato-portal system causes eating: evidence for peripheral glucoreceptors. *Science*, **181**, 858–860.

Oomura, Y. (1987). Regulation of feeding by neural responses to endogenous factors. *News in Physiological Sciences*, **2**, 199–203.

Oomura, Y. and Yoshimatsu, H. (1984). Neural network of glucose monitoring system. *Journal of the Autonomic Nervous System*, **10**, 359–372.

Reilly, F. D., McCuskey, P. A. and McCuskey, R. S. (1978). Intrahepatic distribution of nerves in the rat. *Anatomical Record*, **191**, 55–68.

Ricardo, J. A. and Koh, E. T. (1978). Anatomical evidence of direct projections from the nucleus of the solitary tract to the hypothalamus, amygdala and other forebrain structures in the rat. *Brain Research*, **153**, 1–26.

Ritter, S. and Taylor, J. S. (1990). Vagal sensory neurons are required for lipoprivic but not glucoprivic feeding in rats. *American Journal of Physiology*, **258**, R1395–R1401.

Rogers, R. C. and Hermann, G. E. (1983). Central connections of the hepatic branch of the vagus nerve: a horseradish peroxidase histochemical study. *Journal of the Autonomic Nervous System*, **7**, 165–174.

Sakaguchi, T. and Myaoka, Y. (1981). Reflex motility of the stomach evoked by electrical stimulation of the hepatic vagus nerve. *Experientia*, **37**, 150–151.

Schmitt, M. (1973). Influences of hepatic portal receptors on hypothalamic feeding and satiety centres. *American Journal of Pharmacology*, **225**, 1089–1095.

Skaaring, P. and Bierring, F. (1976). On the intrinsic innervation of normal rat liver. *Cell Tissue Research*, **171**, 141–155.

Skaaring, P. and Bierring, F. (1977). Further evidence for the existence of intralobular nerves in the rat liver. *Cell Tissue Research*, **177**, 287–290.

Strandhoy, J. W. and Williamson, H. E. (1970). Evidence for an hepatic role in the control of sodium excretion. *Proceedings of the Society for Experimental Biology and Medicine*, **133**, 419–422.

Tanaka, K., Inoue, S., Takamura, Y., Jiang, Z. -Y. and Niijima, A. (1986). Arginine sensors in the hepato-portal system and their reflex effects on pancreatic efferents in the rat. *Neuroscience Letters*, **72**, 69–73.

Tsai T. L. (1958). A histological study of the sensory nerves in the liver. *Acta Neurovegetica*, **17**, 354–385.

Weatherford, S. C. and Ritter, S. (1988). Lesion of vagal afferent terminals impairs glucagon-induced suppression of food intake. *Physiology and Behavior*, **43**, 645–650.

Weick, B. G. and Ritter, S. (1986). Dose-related suppression of feeding by intraportal glucagon infusion in rats. *American Journal of Physiology*, **250**, 676–681.

4

Gastrointestinal chemoreception and its behavioural role

NOËL MEÏ

*Laboratoire de Neurobiologie, Neurobiologie de l'Intéroception
CNRS-31, chemin Joseph-Aiguier, 13402 Marseille cedex 09, France*

4.1. Introduction

THE autonomic nervous system (ANS) plays an important role in some of the most elementary mechanisms of nutrition (regulation of digestive tone and the storage and propulsion of nutrients, for example) as well as in more complex mechanisms involved in homeostasis and alimentary behaviour. This has been clearly demonstrated in recent years, thanks to new techniques which have made it possible to study the structure and function of a single neurone of the ANS, and to the fact that the organization and role of this system have recently attracted the attention of clinicians.

The viscera, including the digestive tract, are richly endowed with sensory fibres. This was not the classical view, but is now plain from both morphological and electrophysiological data. These data have led scientists to reconsider concepts of the ANS (especially those concerning the digestive nervous system) and its exact contribution (especially the sensory component) to the regulation of nutrition.

This chapter illustrates current data concerning digestive sensitivity before considering the role of the digestive afferents in nutrition.

4.2. Morphological data

4.2.1. Pathways of sensory digestive fibres

Sensory fibres originating from the digestive tract travel in the nerves of both the parasympathetic ANS (mainly the vagus nerves) and the sympathetic ANS (mainly the splanchnic nerves). However, the parasympathetic system provides the majority of the digestive sensory fibres. In the vagus nerve and its several branches, moreover, the distribution of the vagal fibres differs from that of the splanchnic fibres. The vagus mainly supplies the mucosal and muscular layers of the gastrointestinal tract whereas the splanchnic nerve preferentially innervates the serosal layer.

4.2.2. Type and number of digestive sensory fibres

All the known sorts of fibres are present in the digestive nerves: they can be either myelinated (type A and B) or unmyelinated (type C). However, quantitative ultrastructural studies have revealed a preponderance of sensory fibres in these nerves. They were not identified previously because unmyelinated fibres could not be properly observed under light microscopy (Meï *et al.*, 1981).

Histochemical, immunochemical and autoradiographic methods, in addition to electron-microscopy, have made it possible to study the visceral innervation in great detail. It is possible, for example, to observe a single neurone (a cell body or a central or peripheral ending). Despite the many possibilities offered by the large range of techniques available, difficulties are still encountered however when studying visceral sensory innervation. This is because of its peculiarities such as the smallness of the fibres, the diverse locations of the cell bodies of fibres innervating the same viscera and the wide dispersion of the peripheral and central endings.

4.2.3. Peripheral and central endings

In both the digestive tract and the central nervous system, the endings of sensory fibres are difficult to locate exactly, because they are so widely dispersed. Moreover, at the peripheral level, the sensory endings resemble the endings of motor neurones and intrinsic neurones. Substantial progress has been made however thanks to the techniques of labelling sensory neurones with substances such as peroxidase (HRP-WGA) or radioactive leucine in particular (Sato *et al.*, 1986; Clerc and Condamin, 1987).

It has been confirmed that most of the peripheral endings are free of surrounding receptor structures in the viscera including the digestive tract. The structure of enteroceptors appears to be particularly simple as regards the modest number of terminal branches as well as the absence of any special accessory formations (Fig. 4.1A). More complex structures have also been described in the digestive viscera. These include Pacinian corpuscles situated in the mesentery (Fig. 4.1B). Some differentiated enterocytes might play a role in the transduction of signals to nerve terminals running close to the epithelium (Fig. 4.1C; Newson *et al.*, 1982). Alternatively, the nerve terminals surround enteric neurones that might act as accessory cells (Fig. 4.1D; Christensen, 1984).

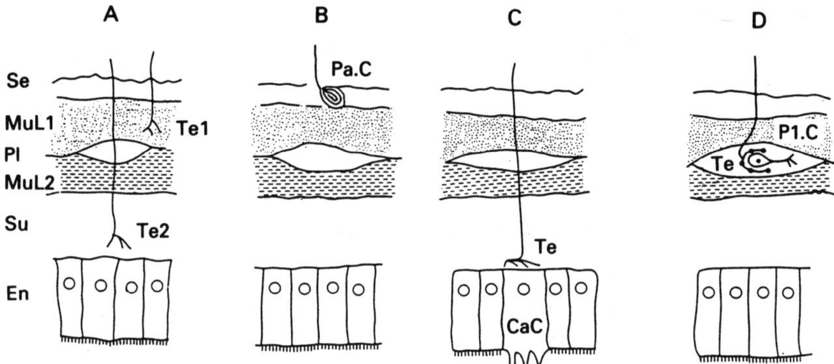

FIG. 4.1.   Structures of sensory receptors in the digestive tract. A. Free endings located within the muscular layers (Te 1) and the mucosa (Te 2). B. Pacinian corpuscle (Pa. C) situated in the serosa. C. Mucosal nerve terminal (Te) ending close to a special enterocyte, a caveolated cell (Ca C). D. Nerve terminal (Te) surrounding a cell (Pl. C) of the myenteric plexus. Se, serosa; Mu L1, circular muscular layer; Pl, myenteric plexus; Mu L2, longitudinal muscular layer; Su, submucosa; En, enterocytes.

4.3. Electrophysiological data

Microelectrode recording has established that the digestive tract gives rise to a wide range of physiological signals. The events accompanying digestion trigger a large number of stimuli concerning the mechanical state of the bowel (local degree of distension or contraction) and the physico-chemical properties of the digestive contents (pH, temperature, osmotic pressure, chemical composition and concentration, etc.) (Fig. 4.2).

4.3.1. Enteroceptor specifity

The problem now faced by neurophysiology is how to identify the type of interoceptors in which these various kinds of messages are elicited. Do receptors respond specifically to only one stimulus? Or are they non-specific or polymodal? These questions are not easy to answer because of the technical difficulties involved in controlling and using intestinal stimuli, especially the chemical ones (see Leek, 1972, 1977).

Nevertheless, in the present state of the art, distinct types of receptor function seem to be present in the digestive tract, including several types of mechanoreceptors, chemoreceptors and thermoreceptors (Fig. 4.3; Meï, 1983, 1985). However, the existence also of truly polymodal receptors has been clearly demonstrated (Andrews and Andrews, 1971; Davison, 1972).

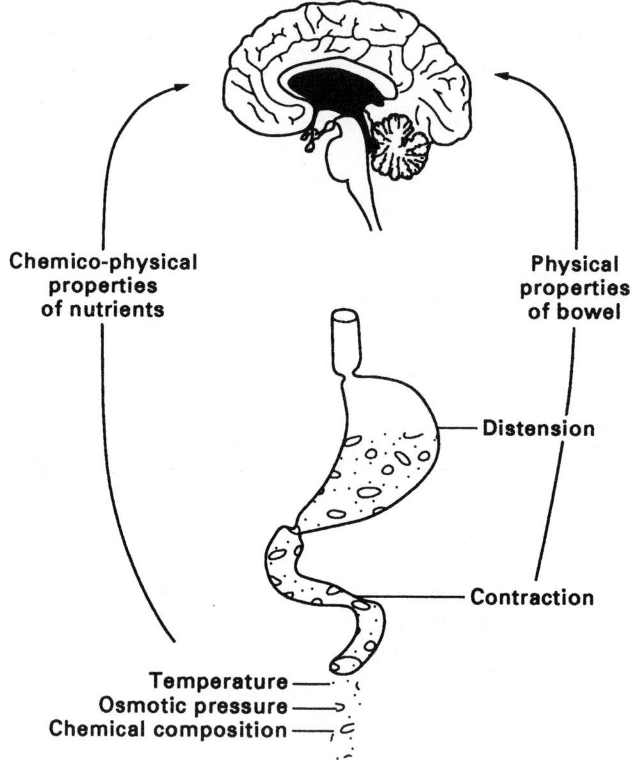

FIG. 4.2.   Complexity of sensory messages provided to CNS by the digestive tract. The two known types of information involve the chemico-physical properties of nutrients (temperature, osmotic pressure, chemical composition etc.) and the physical state of bowel (degree of distension, contraction etc.).

4.3.2. Digestive signal transduction

The function of a receptor was usually thought to be determined by its structure. This idea cannot be applied to the viscera, however, since these tissues mostly contain simple, free endings. The absence of morphological differentiation means that other explanations need to be found for the functional specialization of enteroceptors (Fig. 4.4). The only available hypothesis so far is that the nerve terminals may be endowed with membrane receptors capable of binding specifically one stimulating molecule among numerous sugars, amino acids and fatty acids, for example (Fig. 4.4c).

It is probable that this mechanism is not the only one involved, however. For example, deformation of the villus due to the absorption or secretion of water may possibly be responsible for the activation of osmosensitive receptors (Fig. 4.4b).

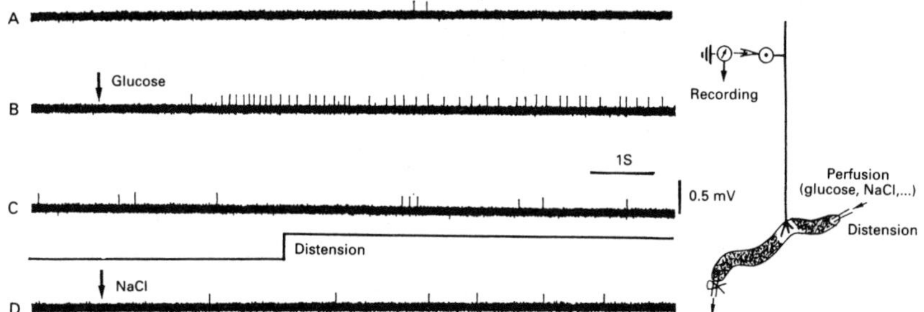

FIG. 4.3.   Example showing the specificity of a intestinal vagal glucoreceptors. The unit is recorded in the nodose ganglion by means of a glass microelectrode. A. Control activity. B. Response to a glucose solution (500 mOsmoles); the arrow indicates the beginning of perfusion. C. Distension 20 minutes later, (50 ml isotonic NaCl) does not evoke clear response in the receptor. D. Perfusion with NaCl solution (500 mOsmoles as in B) does not induce activation in the receptor. The arrow indicates the beginning of perfusion.

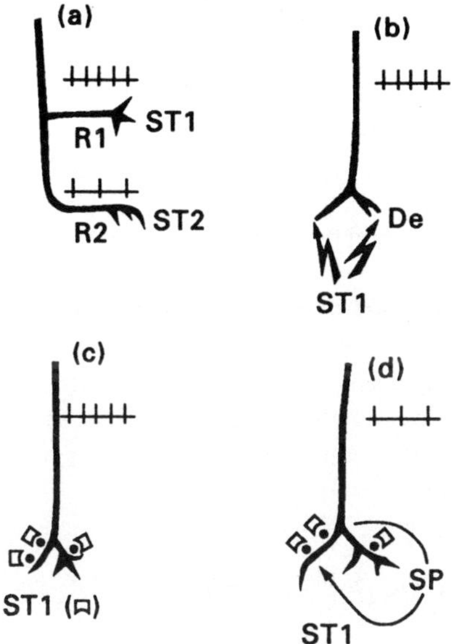

FIG. 4.4.   The mechanisms involved in transduction of sensory digestive signals. a. One fibre might give off two terminals (R1 and R2) sensitive to different stimuli (St. 1 and St. 2). b. The receptor (r) might be activated by deformation of tissue (De, arrows) during digestion. c. The nerve terminal is endowed with membrane receptors (●) capable of binding electively the stimulating substance. St. 1 (□). d. The main mechanism showed in C can be modulated by substances (SP) released during digestion.

In addition, the responses produced by these mechanisms may be modulated by other processes such as the release of neuropeptides by the sensory endings or the neighbouring tissue, which in turn may modify the receptor responses produced by the main mechanism (Fig. 4.4c). This is supported by the existence of special cells (called caveolated cells) in the intestinal epithelium, which have nerve terminals close to their basal membrane (Newson *et al.*, 1982; Fig. 4.1c).

4.4. Functions of digestive afferents

In addition to purely morphological and electrophysiological studies on digestive sensitivity, numerous investigations have been conducted with a view to elucidating the exact role of the digestive afferents. This is difficult to determine, for several reasons.

First, to ascertain the role of any particular kind of digestive receptor, it is necessary to stimulate it specifically. In the visceral area, this is not easy to achieve. For example, perfusion with a substance, intended to stimulate chemoreceptors, may elicit mechanical changes and therefore activate mechanoreceptors.

Secondly, it is not always possible to demonstrate unequivocally the specific role of any one kind of sensory ending in a particular mechanism such as peristalsis. The motor response might be due to other factors, including the triggering of hormone release by the stimuli used.

Thirdly, appropriate techniques are not always available for determine levels of relevant substances such as hormones or neuropeptides in the blood of the species of interest.

Despite these difficulties, it is now clear that visceral afferents including those from the digestive tract are involved in digestive motility, stabilizing the internal environment and alimentary behaviour. Each type of afferent (mechanical, chemical and thermal) is not involved in just one particular function (see Table 4.1). For example, the intestinal glucoreceptors intervene both in motor coordination and insulin release. On the other hand, all the known intestinal receptors are involved in the nervous regulation of gastric emptying, which suggests that this function plays an important part in digestion.

4.4.1. *Control of digestive motility and secretion*

Enteroceptors are involved in various ways in the control of peristalsis in the oesophagus and intestines and in the gastric storage and evacuation of nutrients (Table 4.1). In addition, the afferents participate in controlling the muscular tone of the digestive tract, as well as the migrating myographic complexes (Laplace, 1984). Moreover, they contribute to coordinating the motor activity of different parts of the digestive tract, through various excitatory or inhibitory reflexes. For example, all the known intestinal receptors (mechanoreceptors, thermoreceptors, osmosensitive receptors) trigger different inhibitory intestino-gastric reflexes, the function of which is to slow down gastric evacuation in keeping with the absorptive capacity of the small intestine. Another example is that the

TABLE 4.1 *Main possible roles of gastrointestinal receptors*

Receptor type	Roles
Mechanoreceptors	Maintaining muscular tone and migrating myoelectric complexes, modulation of peristaltic activity, coordination of gastrointestinal motility, control of exocrine secretions, alimentary behaviour
Chemoreceptors	Control of gastric emptying, endocrine and exocrine secretions, intestinal absorption, local circulation, storage of nutrients, alimentary behaviour
Osmosensitive receptors	Control of gastric emptying, diuresis
Thermoreceptors	Control of gastric emptying, thermoregulation, alimentary behaviour

arrival of food in the stomach stimulates the emptying of the colon through excitatory gastro-colic reflexes.

The digestive receptors are also involved in the regulation of exocrine secretions. For example, distension of the stomach elicits an increase in acid secretion through gastric vagal reflexes from the muscular mechanoreceptors.

4.4.2. *Control of absorption*

For some years (Sjövall *et al.*, 1983), the possibility has been considered that the digestive afferents might control the passage of nutrients or the secretion of water and electrolytes within the upper intestine. On the other hand, the central nervous system is involved since it can secondarily modify absorption rates by regulating motor and vasomotor activities.

4.4.3. *Control of homeostasis*

The digestive afferents contribute to maintaining stability of the internal environment such as via regulation of the blood levels of insulin or glucose. Intestinal glucoreceptors trigger a rapid release of insulin which precedes secretion due to the arrival of glucose in the pancreas. This is a true reflex which is mainly mediated by the vagal nerves (Meï *et al.*, 1981).

Intestinal glucoreceptors yield precise preabsorptive information about the availability of glucose. Integrating this information, together with that from other visceral receptors (hepatic, oral, etc.), hypothalamic output might therefore be adapted to peripheral requirements (Niijima and Meï, 1987). Recently, it was suggested (Melone, 1989) that the digestive afferents originating from the receptors sensitive to lipids might also intervene in the peripheral nerve output which controls the metabolism of adipocytes (Takahashi and Shimazu, 1982).

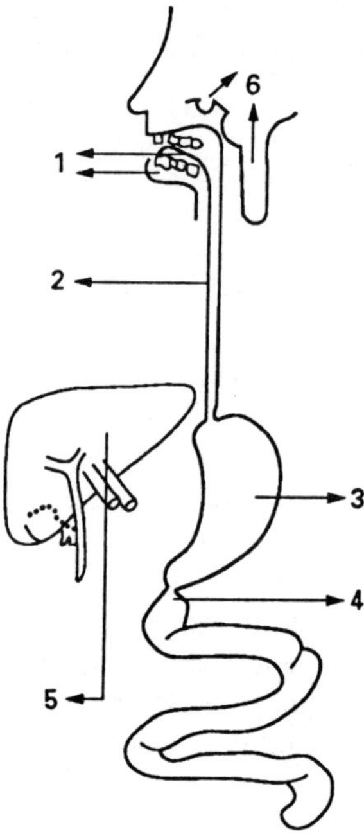

FIG. 4.5.   The nervous signals involved in regulation of ingestion. Some signals arise from the digestive tract. They include: 1: mouth (periodontal receptors and gustatory receptors); 2: oesophagus; 3: stomach; 4: intestine. Others come from the liver: 5 and from the CNS (hypothalamus, medulla): 6.

4.4.4. *Elaboration of alimentary behaviour*

Alimentary behaviour requires the participation of numerous physiologically complex mechanisms (Le Magnen, 1983). Several lines of evidence indicate that internal preabsorptive signals inform the relevant parts of the central nervous system about the state of the digestive tract. As mentioned above, the sensory neural equipment is perfectly able to provide information of this kind about both the mechanical state of the bowel and the physico-chemical properties of nutrients before and as they are digested and absorbed. Nevertheless, the relative importance of signals from different locations (stomach, duodenum, jejunum, ileum) and receptor types (mechanical, thermal or chemical) in different species has yet to be elucidated.

In the opinion of some authors, the main signals arise from the stomach, whereas others have claimed that the small intestine is the main source of internal information (see McHugh and Moran, 1985). From present information, it seems likely that all these signals are involved (Fig. 4.5).

The central nervous system may intervene in modulating behaviourally important signals through various other mechanisms. For example, neural control of gastric emptying is thought to play an important role in food intake by altering the delivery of nutrients to intestinal digestion and absorption (Booth, 1978). Neural participation in blood glucose and insulin levels may be directly involved in triggering or stopping food intake (Niijima, 1975; Niijima and Meï, 1987). Vagal information on the release of gastro-intestinal hormones such as CCK which are thought to be specific signals of satiety (Silver *et al.*, 1989) also influences ingestion.

Digestive afferents play a part in association with the cephalic signals in short-term control of food intake (Nicolaïdis, 1980). Digestive sensory signals can contribute to a decrease in the pleasantness of eating foods and thereby suppress intake (Cabanac and Fantino, 1977). For mechanisms involving conditioning and other forms of learning, information originating from digestive sensory endings is necessary to induce changed responses to gastrointestinal association with oral signals to adjust subsequent acceptance. It has been proposed that duodenal mechano- or chemoreceptors respond to concentrated carbohydrates by aversive conditioning of dietary flavours and mild distension (Booth and Davis, 1973). Recently, special attention has been paid to the vagal and hepatic afferents (Langhans *et al.*, 1988; Martin *et al.*, 1978) which seem to be able to modify aversive conditioning behaviour.

References

Andrews, J. H. and Andrews, H. H. (1971). Receptors activated by acid in the duodenum wall of rabbits. *Quarterly Journal of Experimental Physiology*, **56**, 221–230.

Booth, D. A. (1978). Prediction of feeding behaviour from energy flows in the rat. In D. A. Booth (Ed.), *Hunger Models* (pp. 227–278). London: Academic Press.

Booth, D. A. and Davis, J. D. (1973). Gastrointestinal factors in the acquisition of oral sensory control of satiation. *Physiology and Behavior*, **11**, 23–29.

Cabanac, M. and Fantino, M. (1977). Origin of olfacto-gustatory alliesthesia: intestinal sensitivity to carbohydrate concentration. *Physiology and Behavior*, **18**, 1039–1045.

Christensen, J. (1984). Origin of sensation in the esophagus. *American Journal of Physiology*, **246**, G221–G225.

Clerc, N. and Condamin, M. (1987). Selective labeling of vagal sensory nerve fibers in the lower esophageal sphincter with anterogradely transported WGA-HRP. *Brain Research*, **424**, 216–224.

Davison, J. S. (1972). Response of single vagal afferent fibres to mechanical and chemical stimulation of the gastric and duodenal mucosa in cats. *Quarterly Journal of Experimental Physiology*, **57**, 405–416.

Langhans, W., Kunz, U. and Scharrer, E. (1988). Hepatic vagotomy increases consumption of a novel-tasting diet presented immediately after surgery. *Physiology and Behavior*, **46**, 671–678.

Laplace, J. P. (1984). Motricité de l'intestin grêle: organisation, régulation et fonctions. Quinze ans de recherches sur les complexes migrants. *Reproduction Nutrition Développement*, **24**, 707–765.

Le Magnen, J. (1983). Body energy balance and food intake: a neuroendocrine regulatory mechanism. *Physiological Reviews*, **63**, 314–386.

Leek, B. F. (1972). Abdominal visceral receptors. In E. Neil (Ed.), *Handbook of Sensory Physiology. Enteroceptors* (pp. 113–160). Berlin: Springer-Verlag.

Leek, B. F. (1977). Abdominal and pelvic visceral receptors. *British Medical Bulletin*, **33**, 163–168.

Martin, J. R., Cheng, F. Y. and Novin, D. (1978). Acquisition of learned taste aversion following bilateral subdiaphragmatic vagotomy in rats. *Physiology and Behavior*, **21**, 13–17.

McHugh, J. P. R. and Moran, T. H. (1985). The stomach: A conception of its dynamic role in satiety. In J. M. Sprague and A. N. Epstein (Eds.), *Progress in Psychobiology and Physiological Psychology*, Vol. 11 (pp. 197–232) New York: Academic Press.

Meï, N. (1983). Sensory structures in viscera. In D. Ottoson (Ed.), *Progress in Sensory Physiology* (pp. 1–42). Berlin: Springer-Verlag.

Meï, N. (1985). Intestinal chemosensitivity. *Physiological Reviews* **65**, 211–237.

Meï, N. Arlhac, A. and Boyer, A. (1981). Nervous regulation of insulin release by the intestinal vagal glucoreceptors. *Journal of the Autonomic Nervous System*, **4**, 351–363.

Meï, N., Condamin, M. and Boyer, A. (1980). The composition of the vagus nerve of the cat. *Cell and Tissue Research*, **209**, 423–431.

Melone, J. (1989). Vagal receptors sensitive to lipids in the small intestine of the cat. *Journal of the Autonomic Nervous System*, **17**, 331–341.

Newson, B. H., Ahlman, A., Dahlstrom and Nyhus, L. M. (1982). Ultrastructural observations in the rat ileal mucosa of possible epithelial "taste cells" and submucosal sensory neurons. *Acta Physiologica Scandinavica*, **114**, 161–164.

Nicolaïdis, S. (1980). Hypothalamic convergence of external and internal stimulation leading to early ingestive and metabolic responses. *Brain Research Bulletin*, **5**, Suppl. 4, 97–101.

Niijima, A. (1975). Studies on the nervous regulation mechanism of blood sugar levels. *Pharmacology Biochemistry and Behavior*, **3**, Suppl. 1, 139–143.

Niijima, A. and Meï, N. (1987). Glucose sensors in viscera and control of blood glucose level. *News in Physiological Sciences*, **2**, 164–167.

Sato, M., Yoshizaki, K. and Koyano, H. (1986). Axonal transport of radiolabeled materials in afferent fibers of rabbit vagus nerve following supranodose vagotomy. *Japanese Journal of Physiology*, **36**, 225–230.

Silver, A. J., Flood, J. F., Song, A. M. and Morley, J. E. (1989). Evidence for a physiological role for CCK in the regulation of food intake in mice. *American Journal of Physiology*, **256**, R646–652.

Sjövall, H. S., Redfors, S., Hallbäck, D. A., Eklund, S., Joodal, M. and Lundgren, O. (1983). The effect of splanchnic nerve stimulation on blood flow distribution, villous tissue osmolarity and fluid and electrolyte transport in the small intestine of the cat. *Acta Physiologica Scandinavica*, **117**, 359–365.

Takahashi, A. and Shimazu, T. (1982). Hypothalamic regulation of lipid metabolism in the rat: effect of hypothalamic stimulation of lipogenesis. *Journal of the Autonomic Nervous System*, **6**, 225–235.

5

Gastrointestinal mechanoreception in the control of ingestion

L. ASHLEY BLACKSHAW AND DAVID GRUNDY

Department of Biomedical Science, University of Sheffield, Western Bank, Sheffield S10 2TN, England

5.1. Introduction

ONE of the most obvious ways by which we become aware of the presence of food in the gastrointestinal (GI) tract is by its physical presence, hence the term "fullness". However, the sensation of fullness may also depend on the nutrient density of a meal, which is detected by chemoreceptors and is discussed in detail elsewhere in this volume (Meï, Chapter 4). This chapter will concentrate on the role of gastrointestinal mechanoreceptors in the control of ingestion.

Mechanical stimulation of visceral receptors occurs shortly after the visual, olfactory, gustatory and tactile detection of ingested food, making it one of the primary pre-absorptive signals of entry of food into the body. Its function is to prevent overloading of the digestive capacities of the stomach and small intestine, and as such is a powerful inhibitor of food intake. The inhibition of ingestion by distension of the stomach may be used as a therapeutic tool in the treatment of obesity, which is achieved by the surgical placement of an intragastric balloon (Nieben and Harboe, 1982). Balloons have been used extensively for this purpose over the past decade, with some success at least in the short term (Mathus-Vliegen *et al.*, 1990).

A problem associated with the study of gastrointestinal mechanoreception which has plagued many investigators (including the authors) is the occurrence of powerful reflex changes in motor activity of the gut in response to mechanoreceptor stimulation. Reflexes may take place within the confines of a single organ, such as the stomach, or between structures several metres apart such as the stomach and colon. Local mechanisms such as enteric nervous reflexes and paracrine influences may provide the first level of control, with systemic hormones and reflexes through the CNS being involved at a higher level. The functional changes in GI activity resulting from any of these influences will inevitably lead to an altered afferent input to the CNS and so one is presented with a "chicken-and-egg" situation. In other words, the sensory input to the CNS may be a direct result of a stimulus but, as such, that may lead to a reflex. Alternatively, the afferent signal may be entirely the consequence of a gastrointestinal effector response secondary to local reflexes or hormonal responses to the stimulation of the GI tract.

In addition to autonomic responses, stimulation of visceral mechanoreceptors may give rise to several different types of sensation, for example the fullness of ordinary satiety, bloating, nausea and pain. These depend on the type and location of the sensory ending, the level and pattern of stimulation and the route by which afferent signals reach the central nervous system (CNS). All of the above sensations will result in a reduction or termination of food intake. In human subjects, it is easy to distinguish between a pleasurable feeling of satiety and unpleasant sensations like bloating and nausea. Where one is investigating mechanisms of satiety in an animal, however, painful or nauseating effects are hard to distinguish from effects of normal eating and may interfere or even predominate over that of satiation. Pain is often manifested as stereotyped behavioural responses, but mild pain may be difficult to assess. Nausea is often a prodromal sign of retching and vomiting but, as rodents (a common experimental animal for food-intake studies) do not vomit, it can be difficult to be sure whether a reduction in food intake arises from repleteness or distress.

This chapter focuses first on the physiology and functional anatomy of afferent information from the GI tract to the brain, secondly on the central nervous processing of this information, and thirdly on how mechanoreception relates to the actual motor patterns observed in the gastrointestinal tract during the change from the fasted to the fed state.

5.2. Afferent pathways and receptor types

A complex series of sensory signals is continually generated in different types of afferent fibre from the upper GI tract. These follow the same pathway to the CNS as the autonomic outflow to the GI tract (Fig. 5.1). Nausea and satiety are thought to be mediated through the vagal pathway and pain is mediated by afferent activity following the sympathetic innervation (Cervero, 1988).

5.2.1. The vagus

There are 30,000–50,000 abdominal afferent fibres in the vagi, outnumbering efferent fibres by approximately 9:1 (Andrews, 1986). Vagal afferents terminate peripherally within the smooth muscle and mucosal layers throughout the GI tract, except for the distal colon which is supplied by the pelvic nerves. The endings are unspecialized naked nerve fibres, and as such are difficult to trace to their terminations within the muscle and mucosa due to the limitations of autoradiographic techniques (Sato and Koyano, 1987). From electrophysiolog-

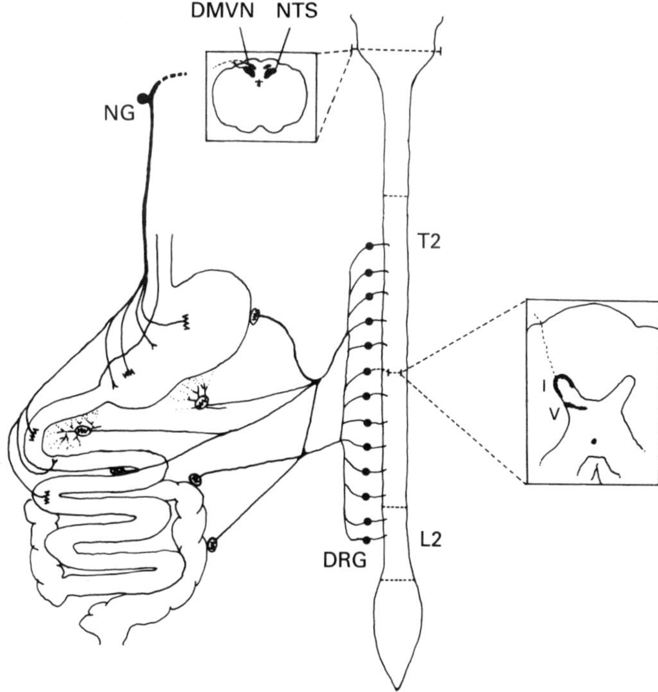

Fig. 5.1.   Schematic summary of central and peripheral terminations of vagal and splanchnic afferent fibres. Upper inset shows a cross-section of the medulla oblongata with major terminations of vagal afferent fibres in nucleus tractus solitarius (NTS) and dorsal motor vagal nucleus (DMVN). Lower inset shows cross-section of spinal cord at T7 with main terminations of splanchnic afferent fibres in Laminae I and V of the dorsal horn grey matter. Cell bodies of afferents are found in NG, nodose ganglia: DRG, dorsal root ganglia. Only right-sided inputs are shown for simplicity.

ical studies, it would appear that the position of endings within the gut wall is important in determining the receptor sensitivity (see later).

The cell bodies of vagal afferent fibres are found in the nodose and jugular ganglia. Central endings terminate in the dorsal medulla oblongata, mainly in the nucleus of the tractus solitarius (NTS), but also in the area postrema and onto vagal efferent neurones in the dorsal motor vagal nucleus (Kalia and Mesulam, 1980; Kalia and Sullivan, 1982; Leslie *et al.*, 1982; Shapiro and Miselis, 1985; Rinaman *et al.*, 1989). From the NTS, information can be disseminated widely to higher areas of the CNS, notably the hypothalamus, which is closely involved with homeostatic mechanisms. A broad range of afferent information is conveyed by the vagi to these areas; however, there are only two main types of vagal sensory fibre, smooth muscle mechanoreceptors and mucosal receptors.

(a) Smooth muscle mechanoreceptors

These endings terminate in the longitudinal and circular muscle layers of the GI tract from the oesophagus to the colon, and show similar characteristics in all species in which they have been studied (for review, see Grundy and Scratcherd, 1989). They normally show an irregular, resting discharge of low frequency, which is related to spontaneous tone or contractile activity in the area of gut containing the receptive field. The activity of these fibres is thus dependent on the mechanical state of the musculature at the ending. They will respond to stretching by distension of the viscera with fluid or a balloon, showing a slowly adapting increase in discharge; the initial discharge frequency is proportional to the intraluminal pressure (Cottrell and Iggo, 1984a; Blackshaw *et al.*, 1987a), which in turn is proportional to the tension in the gut wall — hence the term "tension receptor".

In addition to passive distension, tension receptors will also respond to active contractions, giving rise to a burst of action potentials as each wave of contraction passes over the receptive field. Because of their sensitivity to both stretch and contractions, these endings have been further termed "in-series tension receptors" (Iggo, 1955). This is by analogy with Golgi tendon organs in skeletal muscle, which are, anatomically speaking, true in-series tension receptors. In smooth muscle, the situation is less clear, as it is composed of bundles of closely packed cells which are mechanically coupled. Within these bundles, the individual muscle cells are separated from their neighbours by only tens of nanometres. In such a situation it is difficult to envisage an afferent ending in-series with an individual smooth muscle cell. However, the functional unit is not the cell itself, but the bundle of cells. These are interconnected by collagen fibrils forming laminar intramuscular septa, which are suggested to act as tendons transmitting tension longitudinally from one bundle to the next. The in-series tension receptor in skeletal tendons may therefore have its counterpart in vagal mechanoreceptors if these were located in the connective tissue matrix. In such a location they would respond to muscle tension generated passively during distension and actively during contraction. However, because the intramuscular septa run parallel to muscle cells, these receptors may function as in-series receptors from an in-parallel location.

Regional differences. Although homogeneous in terms of sensitivity to both stretch and contractions, mechanoreceptors in different regions show quite different patterns of activity, because of regional differences in motor functioning. In the gastric antrum, which "pumps" stomach contents into the duodenum, the dominant motor pattern is one of phasic contractile activity. The antrum shows less frequent contractions than the small intestine because of the slower electrical properties of its smooth muscle. Consequently, tension receptors in the

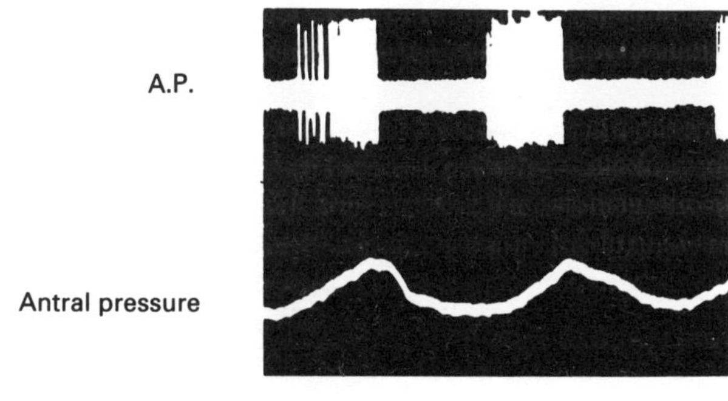

FIG. 5.2.   Discharge of an antral mechanoreceptor recorded from the cervical vagus of the anaesthetized ferret during spontaneously occurring contractions of the antrum (as indicated on the antral pressure trace). A burst of action potentials is generated in the ending as each wave of contraction passes over the receptive field.

gastric antrum and small intestine often show rhythmic bursting at rest, particularly the antrum which shows prominent phasic activity even when empty in the anaesthetized ferret (Andrews *et al.*, 1980; Fig. 5.2). Upon distension, this rhythmic activity becomes more prominent, sometimes superimposed on an increased background of tonic activity. As the antrum is not normally appreciably distended, it may be this tonic discharge that gives rise to sensations of discomfort during gastric stasis and is the means by which vomiting is elicited by antral distension (Grundy *et al.*, 1990).

The fundus and corpus of the stomach are responsible for the reservoir function of the stomach, with the majority of a meal being accommodated in this region, and they show tonic rather than phasic motor activity. As a consequence, mechanoreceptors in the corpus respond primarily to passive distension. However, they will respond to contraction if evoked by electrical vagal stimulation and so they are the same type of receptor as those in the antrum and small intestine. Corpus tension receptors are ideally situated to monitor the volume of ingested material (Fig. 5.3). However, the ability of the stomach to undergo relaxation in response to gastric filling will complicate the behaviour of corpus mechanoreceptors. Under normal circumstances, active relaxation of the corpus will attenuate the rise in pressure (and hence in wall tension) associated with filling. The proximal stomach can therefore accommodate large volumes with rises in intraluminal pressure of just a few centimetres of water (Azpiroz and Malagelada, 1984). Nevertheless, these low distending pressures can have a profound effect on gastrointestinal motility and secretion (see later), indicating that corpus mechanoreceptors are tuned to respond over this narrow pressure range. Moreover, the vagal inhibitory reflexes activated during gastric filling

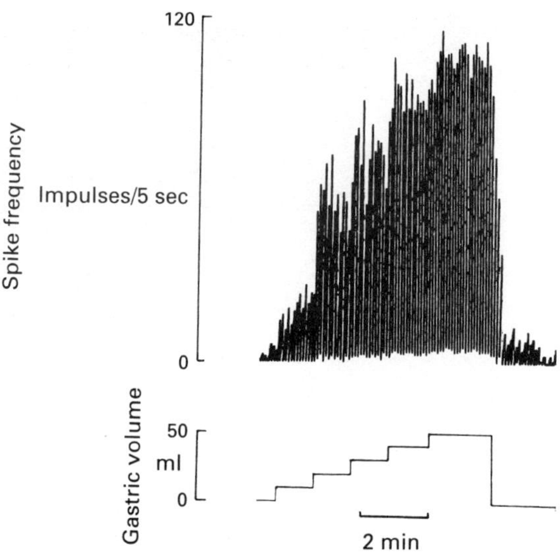

FIG. 5.3.   Integrated record of discharge in a vagal corpus mechanoreceptor (impulses/5 sec) recorded in the anaesthetized ferret on inflation of the stomach in 10-ml steps up to 50 ml. Note the close relationship between discharge frequency and intragastric volume, although this fibre responded primarily to tension.

will subside once food intake has ceased, and the rise in wall tension that follows serves to squeeze gastric contents towards the antral pump. Presumably it will be at this stage that afferent discharge from gastric mechanoreceptors will be at its greatest; this may help to explain the observation that satiety and reflex responses to gastric filling are often delayed. However, as discussed below, this way other receptors may contribute to the mechanisms of satiety.

Afferent input to the CNS. The gastrointestinal tract is endowed with many thousands of tension receptors along its length. So, the brainstem receives a constant stream of information about the mechanical state of the tract, for example the rate, direction and length of propagation of contractions, their strength and the effect they have on the flow of luminal contents. At each stage of digestion, there will be a unique pattern of activity in the gastrointestinal musculature. The volume of afferent information appears sufficient to provide the CNS with considerable detail. It is quite possible, therefore, for satiety to reflect the overall mechanical state of the gut rather than activity in one particular area.

For example, relaxation of the proximal stomach occurs during several aspects of GI function, such as (i) the act of ingestion itself, (ii) feedback inhibition during nutrient delivery to the small intestine, or (iii) nausea and vomiting. In all cases, the relaxation serves to retard the onward delivery of chyme. Clearly, the reflexive, behavioural and sensorial characteristics of corpus filling are not the same under the three conditions. Indeed, under isovolumetric conditions, active relaxation will result in the offloading of tension receptors and a reduction in the

2 sec

FIG. 5.4.   A ferret duodenal vagal mucosal receptor gives a brief burst of firing each time a probe is lightly stroked over the receptive field (bars) otherwise it is silent. This fibre also responded to luminal chemical stimuli, but not to distension of the intact duodenum.

level of afferent discharge from the corpus. Thus, satiety may arise when tension develops in the corpus along with the other aspects of "normal" digestive activity including antral and intestinal peristalsis. If, however, gastric filling occurs as a consequence of the "emetic" pattern of GI motility, which includes retrograde giant contractions to return intestinal contents to the stomach, then a different pattern of afferent activity may arise. The temporal aspects of the afferent signals may also be important. During normal gastric emptying, the gastric volume will decline steadily. A similar volume associated with gastric stasis could be perceived as abnormal.

(b) Mucosal mechanoreceptors

From electrophysiological studies, there appears to be a distinct population of endings in or below the mucosal epithelium of the stomach and small intestine which monitor the physical and chemical nature of luminal contents. Populations of specific nutrient chemoreceptors have been described (see Meï, Chapter 4). However, the majority of mucosal afferents described in the literature possess a mechanical sensitivity in addition to their general chemosensitivity and are known as multimodal afferents. In recent studies mechanical sensitivity in multimodal fibres would appear to predominate over their responsiveness to luminal perfusions (Cottrell and Iggo, 1984b; Blackshaw and Grundy, 1990).

Unlike tension receptors, mucosal receptors do not respond to distension at physiological levels (Clarke and Davison, 1978) but show rapidly adapting responses to probing of the mucosal surface (Fig. 5.4). This exquisite sensitivity to fine tactile stimuli would suggest that they serve a role more as particle detectors rather than as strain gauges (Davison, 1972). Activation of mucosal receptors in the antrum by stroking results in inhibition of corpus tone via a vagal reflex (Andrews and Wood 1988). One might speculate that this reflex would operate to delay gastric emptying by retaining large particles of chyme in the stomach for further trituration before allowing the smaller particles and solution to proceed to the small intestine. In the small intestine, the role of mucosal receptors in the control of food intake is classically viewed as a preabsorptive signal of the chemical composition of a meal (Meï, 1984). However, the sensitivity of these fibres to mechanical events has been largely overlooked and may be important in signalling satiety and generating reflex changes in motility.

Vagal mucosal afferent fibres from the upper gastrointestinal tract can also trigger emesis. Their activation generates a complex sequence of visceral and somatic events which serve to confine the gut contents to the proximal stomach prior to voiding. This has led to another role being proposed for mucosal afferents, to serve as detectors of ingested toxins (Andrews *et al.*, 1990). Thus, mucosal fibres are candidates for the genesis of emesis in addition to reflex control of gastrointestinal function and food intake. Whether specific mucosal receptors are responsible for these different responses is not known. The alternative would be that the pattern and level of afferent discharge is the determining factor.

(c) Influence of gastrointestinal hormones

The role of hormones released from the gut mucosa in the control of ingestion is becoming increasingly apparent. The main candidate for a satiety hormone is cholecystokinin (CCK) although this remains controversial. CCK is released from the small intestinal mucosa on exposure to products of protein or fat digestion in man and other animals (Walsh, 1987; Himeno *et al.*, 1983). When applied exogenously, CCK has a powerful inhibitory effect on food intake (Gibbs *et al.*, 1973). Specific antagonists to CCK, when given alone, result in a significant increase in food intake (Shillabeer and Davison, 1984; Hewson *et al.*, 1988), implying that endogenously released CCK is also effective. CCK also inhibits gastric emptying of a meal (Debas *et al.*, 1975), whereas CCK antagonists accelerate it (Shillabeer and Davison, 1987; Green *et al.*, 1988).

The vagus has long been known to have a major role to play in both satiety and gastric emptying and this is linked with the action of CCK (Smith *et al.*, 1981; Lorenz and Goldman, 1982; Raybould and Tache, 1988). It was postulated that vagal afferent fibres possess CCK receptors on their peripheral endings. The first direct evidence for this arose from studies on rat brainstem neurones by Raybould *et al.* (1985, 1988). They found that NTS neurones always responded in the same way to peripherally administered CCK and to gastric distension (Fig. 5.5). This implies that gastric mechanoreceptors were also directly sensitive to CCK. However, since NTS neurones receive convergent inputs from vagal afferent fibres arising from many different regions of the gastrointestinal tract, there may have been alternative explanations. However, direct recordings of vagal gastric mechanoreceptors have shown that the same fibre may respond to both distension and CCK (Davison and Clarke, 1988). Also, there is preliminary evidence that their sensitivity to CCK (but not distension) is lost following

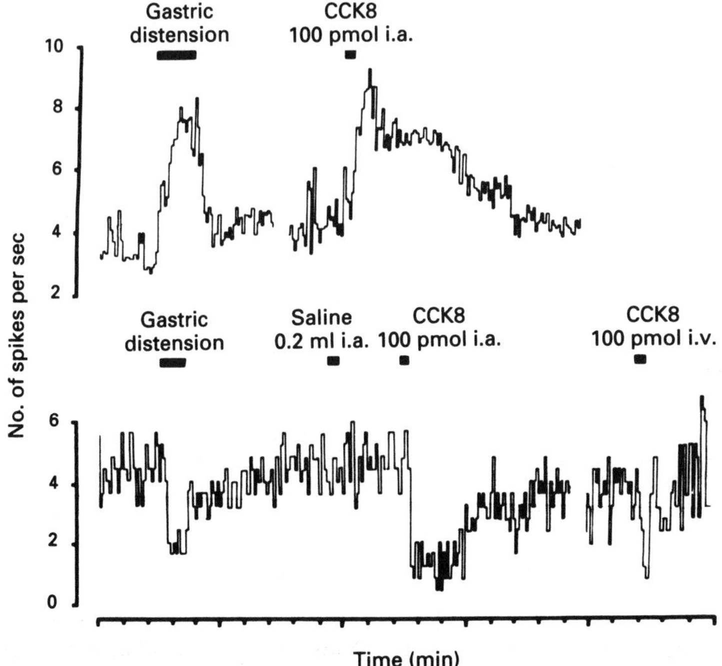

Fig. 5.5.   Effects of CCK and gastric distension on activity of a neurone recorded in the NTS of the rat. Upper trace shows a neurone which was excited by both distension and close intra-arterial CCK-8. Lower trace shows another neurone which is inhibited by both stimuli. Intravenous injections of CCK were without effect. (From Raybould *et al.*, 1985, with kind permission.)

chronic perineural application of capsaicin, which blocks axonal transport in unmyelinated fibres (Raybould and Davison, 1989). Perineural application of capsaicin has a similar effect on brainstem neuronal responses, abolishing those to CCK but not gastric distension (Ritter *et al.*, 1989). In these cases, caudad transport of CCK binding sites was probably disrupted by capsaicin, accounting for the specific effect of the neurotoxin. Capsaicin pretreatment also abolishes the effects of peripheral CCK on gastric emptying (Raybould and Tache, 1988) and food intake (Ritter and Ladenheim, 1985; McCann *et al.*, 1988).

In the rat, therefore, there is good evidence that CCK acts on gastric vagal mechanoreceptors to reduce food intake and to delay gastric emptying by reflex relaxation of the corpus. In human beings, the situation is less clear. If a similar mode of action for CCK and gastric distension were to operate, one would expect that they would cause the manifestation of similar patterns of CNS activity. However, in a recent study by Miaskiewicz *et al.* (1989), it was shown that meal-induced gastric distension caused no change in hypophyseal hormone secretion, whereas CCK administration caused a dose-related increase in

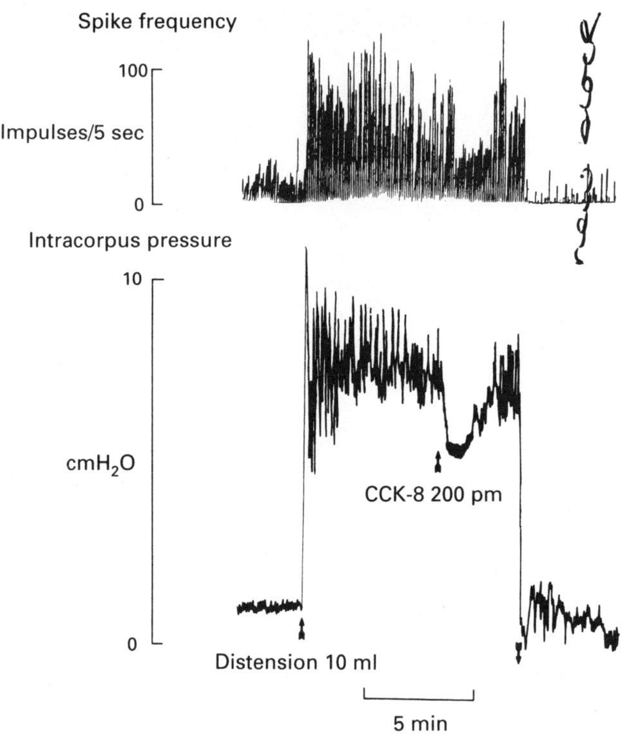

FIG. 5.6. Discharge of a corpus mechanoreceptor (impulses/5sec) recorded from the vagus of the ferret (upper trace), and intracorpus pressure (lower trace). Corpus distension with 10 ml saline causes a slowly-adapting excitation of mechanoreceptor discharge which closely follows intraluminal pressure. CCK-8 (close i.a.) causes immediate relaxation of the corpus for 2–3 minutes, with a corresponding drop in discharge frequency. Discharge and pressure return to control levels over a similar timecourse.

plasma vasopressin levels. This would indicate that there are different mechanisms responsible for effects of the two stimuli in the human species. Furthermore, because vasopressin levels are related to nausea (Rowe *et al.*, 1979), this implies also that exogenous CCK may give rise to activation of brainstem areas responsible for emesis, and therefore cause a reduction in food intake through the sensation of nausea. Indeed, an antiemetic may be antidotal to food-intake suppressor effects of CCK in the rat (Moore and Deutsch, 1985), even though such animals do not vomit. Their sensitivity to the GI pattern of nausea may, however, account for some of the effects of higher doses of CCK on food intake (Verbalis *et al.*, 1986).

In our laboratory we have recently examined the sensitivity of both vagal mucosal receptors and tension receptors to close-intraarterially applied CCK in the ferret (Blackshaw and Grundy, 1990), an animal which may be a better model for man than the rat because of its ability to vomit (Grundy *et al.*, 1990).

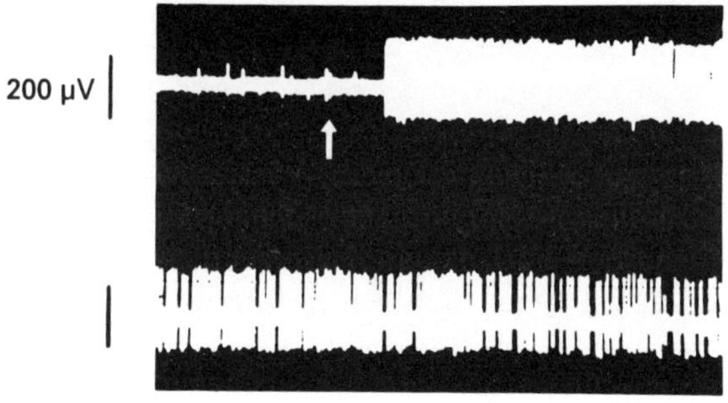

FIG. 5.7.   Effect of 100 pmol CCK-8 injected into the coeliac axis (arrow) on a vagal mucosal mechanoreceptor in the corpus of the ferret. After a latency of approx. 1 second the fibre (previously silent) becomes powerfully excited for > 100 sec. The first 15 sec of the response are shown. This fibre also responded to mucosal stroking and luminally applied hypertonic saline.

Our data show that, as in the rat, CCK causes a profound fall in corpus tone which would retard gastric emptying. However, CCK has no direct effect on the discharge of muscle mechanoreceptors in the stomach or small intestine (Fig. 5.6), as was also found by Cottrell and Iggo (1984a) for sheep duodenal tension receptors. In contrast to tension receptors, mucosal mechanoreceptors in the stomach and duodenum showed powerful afferent responses to CCK (Blackshaw and Grundy, 1990; Fig. 5.7). These responses were unaffected by blockade of motor activity and therefore appear to be due to a direct action of CCK on the mucosal afferent ending. This may indicate that mucosal receptors in the duodenum detect CCK as it is released from epithelial cells, rather than other mechanisms (such as tension receptors) detecting the circulating levels of the hormone. This paracrine mechanism would allow transmission of an accurate picture of CCK release before extensive degradation in the liver has blunted the increase in plasma concentration (Gores *et al.*, 1986). In turn, this would enable rapid signalling of the levels of nutrients which release CCK, albeit through a secondary mechanism. Furthermore, if CCK is infused "downstream" from the gut, into the hepatic-portal vein, its satiating effect is markedly reduced (Greenberg and Smith, 1988). In addition to our electrophysiological studies, vagal sensory fibres have been visualized histologically and traced using autoradiographic techniques to within the mucosal epithelium (Sato and Koyano, 1987). This would support the idea of the mucosal endocrine cells functioning also as transducers of sensory stimuli (see also Fujita *et al.* 1979).

Regardless of the type of vagal afferents activated by CCK, the peripheral CCK receptor has an important role to play in the primary mechanism for signalling satiety from the GI tract. Other gut hormones such as bombesin/GRP and pancreatic glucagon also affect food intake, and their action would appear to be also at least partially via a neural mechanism (Geary and Smith, 1983; Stuckey *et al.*, 1985). There are, however, many other hormones which affect food intake (see Oomura, 1989), and further studies are required to elucidate their precise modes of action and how this may interact with gastrointestinal mechanoreceptors.

5.2.2. *The splanchnic nerves*

Splanchnic afferent fibres have endings throughout the GI tract, and the cell bodies of these fibres are found in the dorsal root ganglia of the spinal cord in segments T2–3 to L2–3. The central terminations are within laminae I and V of the spinal grey matter (Cervero and Connell, 1984; Fig. 5.1). Cells in these areas may also receive an input from somatic structures, but generally this input is only triggered at noxious levels of stimulation (Cervero, 1985). This is the physiological basis for referral of visceral pain to a somatic structure. From the spinal cord, fibres project directly to the thalamus and reticular formation. It is not clear where in the CNS vagal and splanchnic inputs interact before giving rise to sensations and behavioural effects.

The main difference between vagal and splanchnic afferents is in their sensitivity. Electrophysiological studies have largely failed to demonstrate the existence of mucosal receptors such as those found in the vagi, apart from one report of splanchnic glucoreceptors (Perrin *et al.*, 1981). Splanchnic endings are distributed mainly in the serosa and mesentery, often at vascular junctions (Fig. 5.1). One fibre may have up to seven receptive fields, sometimes in different organs (Morrison, 1977). These afferents will respond to visceral distension but not in a manner directly related to intramural tension. Their thresholds for activation are higher than those for vagal mechanoreceptors and they are powerfully excited by stretching of the mesenteric fan or movements of visceral loops against one another. Some splanchnic afferents only respond to noxious levels of stimulation and are sometimes regarded as a separate population to those with a broader dynamic range (see Cervero, 1985 and Janig and Morrison, 1986, for reviews).

The behavioural effects of splanchnic afferent stimulation are commonly associated with pain. However, the sensation of bloating and early satiety in vagotomized individuals is likely to arise from splanchnic mechanoreceptors; whether this is due to plasticity in the sensitivity of the CNS on removal of a vagal input or a normal physiological phenomenon is yet to be resolved. Because of the overlap in the sensitivity of vagal and splanchnic afferents, it is likely that some aspects of sensation are derived from both sources. Fullness, for example, may be the result of the combined input from vagal tension receptors activated by distension and splanchnic afferents activated by tension on the serosal and mesenteric attachments caused by distortion. Activation of splanchnic afferent pathways results in reflex inhibition of GI motility, which by slowing gastric emptying may in turn influence vagal afferent traffic to the CNS. Thus the interaction between vagal and splanchnic systems may occur at several distinct levels.

5.3. Central nervous processing of afferent information

On reaching the first synapse in the CNS, information from the viscera may be disseminated to many other regions of the brain within only a few further synapses. This can make the study of central evoked responses from gastrointestinal stimuli very difficult to interpret. However, controlled ablation studies and microelectrophysiological recordings have indicated that the hypothalamus plays a pivotal role in the integrated control of eating behaviour (Le Magnen, 1988), perhaps in part via modulation of gastrointestinal function.

Single-neurone responses in the NTS integrate afferent information from different areas of the gastrointestinal tract (see earlier). Splanchnic inputs have also been recorded at this level (Barber and Yuan, 1989) and found to be predominantly inhibitory. The resulting activity in NTS neurones may affect neurones in the adjacent dorsal motor vagal nucleus directly, giving rise to vagal reflexes. The information is also transmitted to various nuclei elsewhere in the CNS (e.g. Kobashi and Adachi, 1986).

The lateral (LH) and ventromedial (VMH) nuclei of the hypothalamus have been shown electrophysiologically to receive inputs from the stomach and small intestine (Anand and Pillai, 1967; Jeanningros, 1984) which may be vagal or splanchnic in origin (Jeanningros and Meï, 1980). LH neurones show both excitatory and inhibitory responses to gastric or intestinal distension in the cat (Barone, *et al.*, 1979; Jeanningros, 1984). It appears that LH neurones receive convergent sensory signals of visual, gustatory and olfactory origin, in addition to GI inputs. Thus the firing rate of LH neurons may reflect the overall ingestive behavioural state of the whole animal (Rolls, 1981 and Chapter 9).

The hypothalamus possesses chemoreceptive cells itself (Oomura *et al.*, 1974). It thus provides a good location for the integration of visceral, somatosensory and systemic signals. How much processing visceral afferent information has undergone before reaching the hypothalamus is not known. It would be interesting to identify, for example, the site of convergence of vagal information from mucosal receptors and from mechanoreceptors. This may be important in determining which features of the afferent input are involved in autonomic reflexes and which are involved with behavioural responses. Indeed, electrophysiological studies of vagal efferent responses to different types of gastrointestinal stimuli indicate that mechanoreceptive inputs dominate chemoreceptive inputs in their effectiveness in generating vago-vagal reflexes

(Blackshaw *et al.*, 1987b, 1989). A corollary might be that chemoreceptors are more involved in centrally organized ingestive behaviour than in visceral reflex responses. However, the transmission of afferent information through the brainstem is not fixed but subjected to descending influences from the hypothalamus, with the paraventricular nucleus of particular importance (Rogers and Hermann, 1985a,b). Putative neurotransmitters in these descending pathways such as thyrotropin releasing hormone (TRH) and oxytocin have profound and often opposite effects on neurones in NTS and the dorsal motor nucleus of the vagus (McCann and Rogers, 1990; McCann *et al.*, 1989). This suggests that the hypothalamus not only modulates gastrointestinal function but also may influence ascending input to the hypothalamus analogously to the thalamic influence on inputs to the somatosensory cortex.

5.4. Motor activity of the GI tract in fasted and fed states

In order to gain a clear idea of the picture of gastrointestinal activity that the brain constructs from the enormous volume of afferent information, it is important to understand the effects that ingestion of food has on the gut itself. In the fasted and fed states, the GI tract shows completely different patterns of activity. This section concerns how the transition between the two occurs and how gastrointestinal motility may affect ingestive behaviour and food intake.

5.4.1. Fasted motility

In the interdigestive state, after the upper GI tract has emptied, the stomach and small intestine display a cyclical pattern of activity known as the migrating motor (or myoelectric) complex or MMC (Szurszewski, 1969). In people, this recurs approximately every 90 minutes. It consists of three phases, the first being quiescent, the second consisting of a period of moderate amplitude irregular contractions, and the third comprising a series of high-amplitude, regular peristaltic contractions (Figs. 5.8 and 5.9). This motor complex migrates slowly down the GI tract from the stomach as far as the ileocaecal junction. It serves as an "interdigestive housekeeper" to sweep out the gut of residual food and secretions.

During phase III of the MMC, contractions are of high amplitude and maximal frequency. A phase III is always present somewhere in the gastrointestinal tract. The presence of this contractile pattern in the stomach may correspond to the "hunger contractions" described by Cannon and Washburn (1912). Thus the MMC may also serve to reinforce or even to generate sensations of hunger, such as the epigastric pang, and to motivate ingestion.

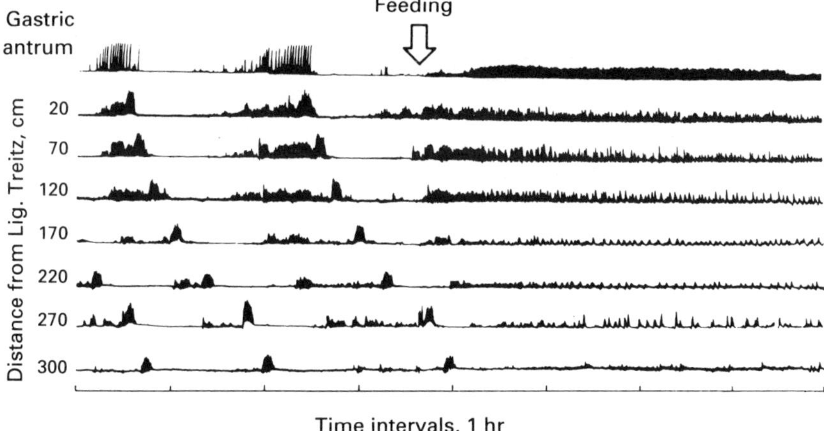

FIG. 5.8.   Eight-hour recording of contractile activity in the GI tract of a conscious dog. Each MMC of contractile activity can be visualized moving slowly down the small intestine with three distinct phases of activity (see text). Upon feeding a continuous pattern of motility appears at all levels and persists for several hours. The ligament of Treitz marks the end of the duodenum. (Reprinted from "Interdigestive motor activity in health and disease", by Itoh and Sekiguchi (1982) from *Scandinavian Journal of Gastroenterology*, by permission of Universitetsforlaget AS/ Norwegian University Pres.

5.4.2. Fed motility

After a short delay following the ingestion of a meal, the MMC is disrupted and replaced by continuous irregular motor activity in both stomach and small intestine (Fig. 5.8). The change in volume and composition of intraluminal contents following the ingestion of a meal stimulates a variety of reflexes and releases a number of hormones which initiate and maintain the post-prandial motor and secretory responses. The situation for motility is analogous to that for gastric secretion with stimuli arising at several different levels, giving cephalic, gastric and intestinal phases. This pattern of motility can be evoked by gastric distension at remarkably low levels (Code and Marlett, 1975; Azpiroz and Malagelada, 1984; Bull *et al.*, 1987).

During and after ingestion, motility in different regions of the gastrointestinal tract is highly coordinated. As food is propelled down the oesophagus, the resulting distension results in a vagal reflex which relaxes the muscle of the proximal region of the stomach (corpus and fundus). This is known as receptive relaxation (Cannon and Lieb, 1911). It allows the stomach to accommodate large volumes of food with minimal increases in pressure. Reflex relaxation can also occur in response to gastric distension, which is similarly mediated by the vagus (Abrahamsson, 1973) and is known as accommodation. The proximal stomach therefore behaves as a reservoir and hopper for food, delivering

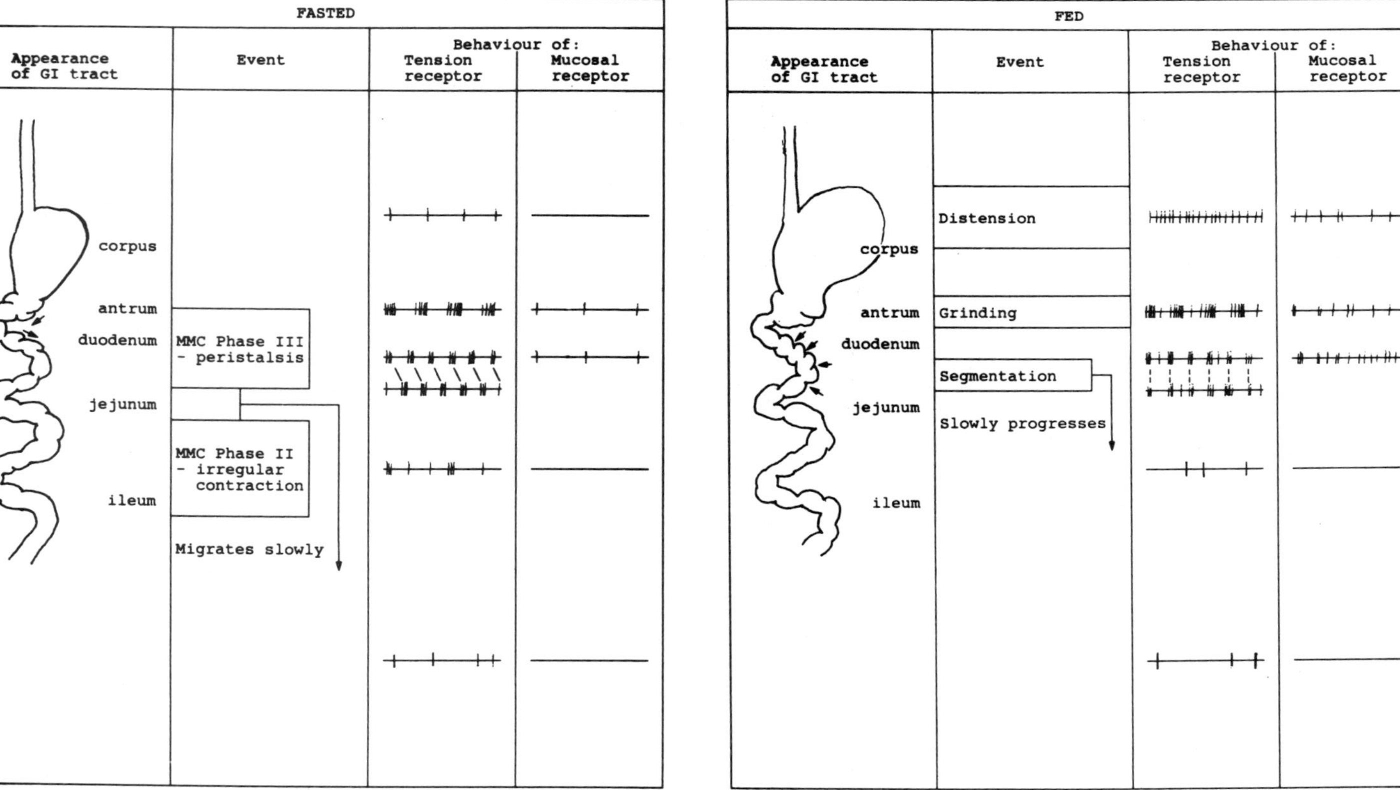

Fig. 5.9.   Schematic representation of motor patterns in the gastrointestinal tract in fasted and fed states, together with the suggested behaviour of tension and mucosal receptors in different regions. During fasting, the main stimulus to mucosal receptors is probably mechanical due to mucosal surfaces touching during phase III of the MMC, whereas after feeding they receive both mechanical and chemical stimuli due to contact with food and secretion (which is only present in the proximal GI tract in this schematic). Tension receptors respond to the level of distension or contractions, and are closely spaced so that peristaltic (fasted) or segmenting (fed) contractile waves can be signalled to the CNS. Splanchnic afferent fibres are not illustrated as they are likely to show similar patterns of weak discharge in both fasted and fed states, unless overdistension

it gradually to the distal stomach where it is ground and emptied into the duodenum.

The ability of the stomach to accommodate is severely disrupted in patients who have undergone proximal vagotomy for the treatment of peptic ulcer disease. These patients therefore perceive satiety earlier during a meal and feel bloated for some time afterwards (Kelly, 1984). These symptoms highlight the role of mechanoreception from the stomach in the control of food intake. However, because of the surgical interruption in the vagal afferent pathway, it is the splanchnic afferents that are implicated. In addition, because of the accelerated emptying of liquids following proximal vagotomy, intestinal stimuli may play a role, either directly or because of their inhibitory effects on gastric emptying.

This example serves to illustrate how in studies of food intake it is important to reconcile the intake-suppressor potency of treatments with their effects on gastric emptying. The powerful inhibition of food intake by CCK is another case in point, because CCK also reduces gastric emptying by relaxing the proximal stomach and contracting the pylorus and duodenum (e.g. Scheurer *et al.*, 1983). The effect on emptying was proposed to be the only reason why CCK causes satiety (Moran and McHugh, 1982) but, as discussed above, the situation is now agreed to be more complex, given CCK's direct action on vagal afferents.

In the small intestine, segmenting contractions and short propagating waves facilitate absorption by ensuring complete mixing of chyme with digestive enzymes and its contact with a large surface area of epithelium (Fig. 5.9). Mechanical distension and the action of nutrients on the wall of the small intestine have a marked effect on gastrointestinal motility and generally result in a slowing of gastric emptying and intestinal transit. Sensations ranging from bloating through discomfort to pain can be envisaged with different degrees of upper intestinal distension and could contribute to an inhibition in food intake.

The same might be said for the colon. However, filling and emptying of the small intestine occurs within the normal interval between meals, with resumption of the MMC marking the return to the fasting pattern of motor activity. Transit through the colon is very much slower. Thus, colonic mechanoreceptors are probably not important in the control of ingestion unless distension reaches levels causing discomfort.

5.5. Conclusions

The mechanical state of the gastrointestinal tract has an important role, not only in ensuring appropriate digestion of food, but also in the genesis of satiety and therefore in ingestive behaviour. The brain receives a complex picture of contraction, relaxation, distension and distortion of the gut musculature, along with a wealth of information on the consistency and composition of its contents from the many thousands of afferent fibres throughout the gastrointestinal tract. There is therefore a great deal of decoding of these signals to be done in order to initiate the appropriate behavioural and reflex responses. It is unlikely that a single population of sensory fibres gives rise to the diverse sensations and behavioural responses emanating from the GI tract. A more likely explanation is that the input from different receptors in different regions forms a code which the CNS can interpret. The investigation of how the various afferent inputs from the gastrointestinal tract are integrated in the central nervous system is an essential part of the study of the mechanisms controlling food intake.

References

Abrahamsson, H. (1973). Studies on the inhibitory nervous control of gastric motility. *Acta Physiologica Scandinavica*, Suppl. 390, 38.

Anand, B. K. and Pillai, R. V. (1967). Activity of single neurons in the hypothalamic feeding centres: effect of gastric distension. *Journal of Physiology*, **192**, 63–67.

Andrews, P. L. R. (1986). Vagal afferent innervation of the gastrointestinal tract. In (Eds.), F. Cervero and J. F. B. Morrison, *Visceral Sensation, Progress in Brain Research*, Vol. 67 (pp. 65–86). Elsevier: Amsterdam.

Andrews, P. L. R., Davis, C. J., Bingham, S., Davidson, H. I. M., Hawthorn, J. and Maskell, L. (1990). The abdominal visceral innervation and the emetic reflex: pathways, pharmacology, and plasticity. *Canadian Journal of Physiology and Pharmacology*, **68**, 325–345.

Andrews, P. L. R., Grundy, D. and Scratcherd, T. (1980). Vagal afferent discharge from mechanoreceptors in different regions of the ferret stomach. *Journal of Physiology*, **298**, 513–524.

Andrews, P. L. R. and Wood, K. L. (1988). Vagally mediated gastric motor and emetic reflexes evoked by stimulation of the antral mucosa in anaesthetized ferrets. *Journal of Physiology*, **395**, 1–16.

Azpiroz, F. and Malagelada, J-R. (1984). Pressure activity patterns in the canine proximal stomach: response to distension. *American Journal of Physiology*, **247**, G265–G272.

Barber, W. D. and Yuan, C-S. (1989). Gastric-vagal-splanchnic interactions in the brainstem of the cat. *Brain Research*, **487**, 1–8.

Barone, F. C., Wayner, M. J., Weiss, C. S. and Almli, C. R. (1979). Effects of intragastric water infusion and gastric distension on hypothalamic neuronal activity. *Brain Research Bulletin*, **4**, 267–282.

Blackshaw, L. A. and Grundy, D. (1989). Responses of vagal efferent fibres to stimulation of gastric mechano- and chemoreceptors in the anaesthetized ferret. *Journal of the Autonomic Nervous System*, **27**, 39–45.

Blackshaw, L. A. and Grundy, D. (1990). Effects of cholecystokinin on two classes of gastroduodenal vagal afferent fibre. *Journal of the Autonomic Nervous System*, **31**, 191–202.

Blackshaw, L. A., Grundy, D. and Scratcherd, T. (1987a). Vagal afferent discharge from gastric mechanoreceptors during contraction and relaxation of the ferret corpus. *Journal of the Autonomic Nervous System*, **18**, 19–24.

Blackshaw, L. A., Grundy, D. and Scratcherd, T. (1987b). Involvement of gastrointestinal mechano- and intestinal chemoreceptors in vagal reflexes: an electrophysiological study. *Journal of the Autonomic Nervous System*, **18**, 225–234.

Bull, J. S., Grundy, D. and Scratcherd T. (1987). Disruption of the jejunal migrating motor complex by gastric distension and feeding in the dog. *Journal of Physiology*, **394**, 381–392.

Cannon, W. B. and Lieb, C. M. (1911). The receptive relaxation of the stomach. *American Journal of Physiology*, **29**, 270–273.

Cannon, W. B. and Washburn, A. L. (1912). An explanation of hunger. *American Journal of Physiology*, **29**, 441–454.

Cervero, F. (1985). Visceral nociception: peripheral and central aspects of visceral nociceptive systems. *Philosophical Transactions of the Royal Society of London, Series B*, **308**, 325–337.

Cervero, F. (1988). Neurophysiology of gastrointestinal pain. *Baillière's Clinical Gastrenterology*, **2**, 183–200.

Cervero, F. and Connell, L. A. (1984). Distribution of somatic and visceral primary afferent fibres within the thoracic spinal cord of the cat. *Journal of Comparative Neurology*, **230**, 88–98.

Clarke, G. D. and Davison, J. S. (1978). Mucosal receptors in the gastric antrum and small intestine of the rat with afferent fibres in the cervical vagus. *Journal of Physiology*, **284**, 55–67.

Code, C. F. and Marlett, J. A. (1975). The interdigestive myoelectrical complex of the stomach and small bowel of dogs. *Journal of Physiology*, **246**, 289–309.

Cottrell, D. F. and Iggo, A. (1984a). The responses of duodenal tension receptors in sheep to pentagastrin, cholecystokinin, and some other drugs. *Journal of Physiology*, **354**, 477–495.

Cottrell, D. F. and Iggo, A. (1984b). Mucosal enteroceptors with vagal afferent fibres in the proximal duodenum of sheep. *Journal of Physiology*, **354**, 497–522.

Davison, J. S. (1972). Responses of single vagal afferent fibres to mechanical and chemical stimulation of the gastric and duodenal mucosal in cats. *Quarterly Journal of Experimental Physiology*, **57**, 405–416.

Davison, J. S. and Clarke, G. D. (1988). Mechanical properties and sensitivity to CCK of vagal gastric slowly adapting mechanoreceptors. *American Journal of Physiology*, **255**, G55–G61.

Debas, H. T., Farooq, O. and Grossman, M. I. (1975). Inhibition of gastric emptying is a physiological action of cholecystokinin. *Gastroenterology*, **68**, 1211–1217.

Fujita, T., Kobayashi, S., Muraki, S., Sato, K. and Shimoji, K. (1979). Gut endocrine cells as chemoreceptors. In A. Miyoshi and M. I. Grossman (Eds.), *Gut Peptides: Secretion, Function and Clinical Aspects* (pp. 47–52). Kodanasha: Tokyo.

Geary, N. and Smith, G. P. (1983). Selective hepatic vagotomy blocks pancreatic glucagon's satiety effect. *Physiology and Behavior*, **31**, 391–394.

Gibbs, J., Young, R. C. and Smith, G. P. (1973). Cholecystokinin decreases food intake in rats. *Journal of Comparative and Physiological Psychology*, **84**, 488–495.

Gores, G. J., Larusso, N. F. and Miller, L. J. (1986). Hepatic processing of cholecystokinin peptides. I. Structural specificity and mechanism of hepatic extraction. *American Journal of Physiology*, **250**, G344–G349.

Green, T., Dimaline, R. and Dockray, G. J. (1988). Action of the cholecystokinin antagonist L364,718 on gastric emptying in the rat. *American Journal of Physiology*, **255**, G685–G689.

Greenberg, D. and Smith, G. P. (1988). Hepatic portal infusion reduces the satiating potency of CCK-8. *Physiology and Behavior*, **36**, 535–538.

Grundy, D., Andrews, P. L. R. and Blackshaw, L. A. (1990). Neural correlates of the gastrointestinal motor changes in emesis. In Y. Tache and D.L Wingate (Eds.), *Brain Gut Interactions*, CRC: Boca Raton.

Grundy, D. and Scratcherd, T. (1989). Sensory afferents from the gastrointestinal tract. In J. D. Wood (Ed.), *Handbook of Physiology*, Section 6, Vol. I, Part I (pp. 593–620). American Physiological Society: Bethesda, USA.

Hewson, G., Leighton, G.E., Hill, R.G. and Hughes, J. (1988). The cholecystokinin antagonist L364,718 increases food intake in the rat by attenuation of the action of endogenous cholecystokinin. *British Journal of Pharmacology*, **93**, 79–84.

Himeno, S., Tarui, S., Kanayama, T. Kurokawa, M., Shinomura, Y., Hayashi, C., Tateishi, K., Imagawa, K., Hashimura, E. and Hamaoka, T. (1983). Plasma cholecystokinin responses after ingestion of liquid meal and intraduodenal infusion of fat, amino acids, or hydrochloric acid in man: analysis with region specific radioimmunoassay. *American Journal of Gastroenterology*, **78**, 703–707.

Iggo, A. (1955). Tension receptors in the stomach and the urinary bladder. *Journal of Physiology*, **128**, 593–607.

Itoh, Z. and Sekiguchi, T. (1982) Interdigestive motor activity in health and disease. *Scandinavian Journal of Gastroenterology*, **18**, Suppl. 82, 497–521.

Janig, W. and Morrison, J. F. B. (1986). Functional properties of spinal visceral afferents supplying abdominal and pelvic organs, with special emphasis on visceral nociception.

In F. Cervero and J. F. B. Morrison (Eds.), *Visceral Sensation, Progress in Brain Research*, Vol. 67. (pp. 87–114). Elsevier: Amsterdam.

Jeanningros, R. (1984). Modulation of lateral hypothalamic single unit activity by gastric and intestinal distension. *Journal of the Autonomic Nervous System*, **11**, 1–11.

Jeanningros, R. and Meï, N. (1980). Vagal and splanchnic effects at the level of the ventro-median nucleus of the hypothalamus (VMH) in the cat. *Brain Research*, **185**, 239–251.

Kalia, M. and Mesulam, M. M. (1980). Brain stem projections of sensory and motor components of the vagus complex in the cat: II. Laryngeal, tracheobronchial, pulmonary, cardiac and gastrointestinal branches. *Journal of Comparative Neurology*, **193**, 467–508.

Kalia, M. and Sullivan, J.M. (1982). Brainstem projections of sensory and motor components of the vagus nerve in the rat. *Journal of Comparative Neurology*, **211**, 248–264.

Kelly, K. A. (1984). Effect of gastric surgery on gastric motility and emptying. In L. M. A. Akkermans, A. G. Johnson and N. W. Read (Eds.), *Gastric and Gastroduodenal Motility* (pp. 241–262). Praeger: Eastbourne.

Kobashi, M. and Adachi, A. (1986). Projection of nucleus tractus solitarius units influenced by hepatoportal afferent signal to parabrachial nucleus. *Journal of the Autonomic Nervous System*, **16**, 153–158.

Le Magnen, J. (1988). Control of eating behaviour. *Baillière's Clinical Gastroenterology*, **2**, 169–182.

Leslie, R. A., Gwyn, D. G. and Hopkins, D. A. (1982). The central distribution of the cervical vagus nerve and gastric afferent projections in the rat. *Brain Research Bulletin*, **8**, 37–43.

Lorenz, D. N. and Goldman, S. A. (1982). Vagal mediation of the cholecystokinin satiety effects in rats. *Physiology and Behavior*, **29**, 599–604.

Mathus-Vliegen, E. M. H., Tytgat, G. N. J. and Veldhuyzen-Offermans, E. A. M. L. (1990). Intragastric balloon in the treatment of super morbid obesity. *Gastroenterology*, **99**, 362–369.

McCann, M. J., Hermann, B. E. and Rogers, R. C. (1989). Thyrotropin releasing hormone: effects on identified neurons of the dorsal vagal complex. *Journal of the Autonomic Nervous System*, **26**, 107–112.

McCann, M.J. and Rogers, R.C. (1990). Oxytocin excites gastric-related neurones in rat dorsal vagal complex. *Journal of Physiology*, **428**, 95–108.

McCann, M. J., Verbalis, J. G. and Stricker, E. M. (1988). Capsaicin attenuates multiple responses to cholecystokinin in rats. *Journal of the Autonomic Nervous System*, **23**, 265–272.

Meï, N. (1984). Sensory structures in the viscera. *Progress in Sensory Physiology*, **4**, 1–42.

Miaskiewicz, S. L., Stricker, E. M. and Verbalis, J. G. (1989). Neurohypophyseal secretion in response to cholecystokinin but not meal-induced gastric distention in humans. *Journal of Clinical Endocrinology and Metabolism*, **68**, 837–843.

Moore, B. O. and Deutsch, J. A. (1985). An antiemetic is antidotal to the satiety effects of cholecystokinin, *Nature*, **315**, 321–322.

Moran, T. H. and McHugh, P. R. (1982). Cholecystokinin suppresses food intake by inhibiting gastric emptying. *American Journal of Physiology*, **242**, R491–R497.

Morrison, J. F. B. (1977). The afferent innervation of the gastrointestinal tract. In F. P. Brooks and P. W. Evers, (Eds.), *Nerves and the Gut* (pp. 297–326). CBS: New Jersey.

Nieben, O. G. and Harboe, H. (1982). Intragastric balloon as an artificial bezoar for treatment of obesity, *Lancet*, **1**, 198–199.

Oomura, Y. (1989). Sensing of endogenous chemicals in control of feeding. *Progress in Sensory Physiology*, **9**, 171–191.

Oomura, Y., Ooyama, H., Sugimori, M., Nakamura, T. and Yamada, Y. (1974). Glucose inhibition of the glucose-sensitive neurone in the rat lateral hypothalamus. *Nature*, **247**, 284–286.

Perrin, J., Crousillat, J. and Meï, N. (1981). Assessment of true splanchnic glucoreceptors in the jejuno-ileum of the cat. *Brain Research Bulletin*, **7**, 625–628.

Raybould, H. E. and Davison, J. S. (1989). Perivagal application of capsaicin abolishes the response of vagal gastric mechanoreceptors to cholecystokinin. *Society for Neuroscience Abstracts*, **15**, 973.

Raybould, H. E., Gayton, R. J. and Dockray, G. J. (1985). CNS effects of circulating CCK8:

involvement of brainstem neurons responding to gastric distension. *Brain Research*, **342**, 187–190.

Raybould, H. E., Gayton, R. J. and Dockray, G. J. (1988). Mechanisms of action of peripherally administered cholecystokinin octapeptide on brain stem neurons in the rat. *Journal of Neuroscience*, **8**, 3018–3024.

Raybould, H. E. and Taché, Y. (1988). Cholecystokinin inhibits gastric motility and emptying via a capsaicin-sensitive vagal pathway in rats. *American Journal of Physiology*, **255**, G242–G246.

Rinaman, L., Card, J. P., Schwaber, J. S. and Miselis, R. R. (1989). Ultrastructural demonstration of a gastric monosynaptic vagal circuit in the nucleus of the solitary tract in rat. *Journal of Neuroscience*, **9**, 1985–1996.

Ritter, R. C. and Ladenheim, E. E. (1985). Capsaicin pretreatment attenuates suppression of food intake by cholecystokinin. *American Journal of Physiology*, **248**, R501–R504.

Ritter, R. C., Ritter, S., Ewart, W. R. and Wingate, D. L. (1989). Capsaicin attenuates hindbrain neuron responses to circulating cholecystokinin. *American Journal of Physiology*, **257**, R1162–R1168.

Rogers, R. C. and Hermann, G. E. (1985a). Vagal afferent stimulation-evoked gastric secretion suppressed by paraventricular nucleus lesion. *Journal of the Autonomic Nervous System*, **13**, 191–199.

Rogers, R. C. and Hermann, G. E. (1985b). Gastric-vagal solitary neurones excited by paraventricular nucleus lesion. *Journal of the Autonomic Nervous System*, **14**, 351–362.

Rolls, E. T. (1981). Central nervous mechanisms related to feeding and appetite. *British Medical Bulletin*, **37**, 131–134.

Rowe, J. W., Shelton, R. L., Helderman, H., Vestal, R. E. and Robertson, G. L. (1979). Influence of the emetic reflex on vasopressin release in man. *Kidney International*, **16**, 729–735.

Sato, M. and Koyano, H. (1987). Autoradiographic study on the distribution of vagal afferent nerve fibers in the gastroduodenal wall of the rabbit. *Brain Research*, **400**, 101–109.

Scheurer, U., Varga, L., Drack, E., Burki, H-R and Halter, F. (1983). Measurement of cholecystokinin octapeptide-induced motility of rat antrum, pylorus, and duodenum *in vitro*. *American Journal of Physiology*, **244**, G261–G265.

Shapiro, R. E. and Miselis, R. R. (1985). The central organization of the vagus nerve innervating the stomach of the rat. *Journal of Comparative Neurology*, **238**, 473–488.

Shillabeer, G. and Davison, J. S. (1984). The cholecystokinin antagonist, proglumide, increases food intake in the rat. *Regulatory Peptides*, **8**, 171–176.

Shillabeer, G. and Davison, J. S. (1987). Proglumide, a cholecystokinin antagonist, increases gastric emptying in rats. *American Journal of Physiology*, **252**, R353–R360.

Smith, G. P., Jerome, C., Cushin, B. J., Eterno, R. and Simansky, K. J. (1981). Afferent axons in the abdominal vagus mediate the satiety effect of cholecystokinin in the rat. *American Journal of Physiology*, **249**, R638–R641.

Stuckey, J. A., Gibbs, J. and Smith, G. P. (1985). Neural disconnection of gut from brain blocks bombesin-induced satiety. *Peptides*, **6**, 1249–1252.

Szurszewski, J. H. (1969). A migrating electric complex of the canine small intestine. *American Journal of Physiology*, **217**, 1757–1763.

Verbalis, J. G., McCann, M. J., McHale, C. M. and Stricker, E. H. (1986). Oxytocin secretion in response to cholecystokinin and food: differentiation of nausea from satiety. *Science*, **232**, 1417–1419.

Walsh, J. H. (1987). Gastrointestinal hormones. In L. R. Johnson (Ed.), *Physiology of the Gastrointestinal Tract* (pp. 181–253). Raven: New York.

6

The sense of touch in the control of ingestion

T. MORIMOTO[1] and K. TAKADA[2]

[1]*Department of Oral Physiology and* [2]*Department of Orthodontics, Faculty of Dentistry, Osaka University, Japan*

6.1. Introduction

IN mammals, feeding behaviour can be elicited and terminated by factors originating within the body. However, the decision to ingest food, i.e. whether or not to chew and swallow materials carried into the mouth, is determined by orally sensed properties such as taste, texture and temperature.

This chapter discusses the roles of mechanoreceptors in the control of jaw and tongue movements during chewing, as a known neurophysiological background to tactile perception of foodstuffs.

6.2. Physiological role of oral sensation

We take meals not only for life but also for the pleasure of eating. Each food has its own texture that contributes to its palatability by tactile sensation. For example, a crisp apple may feel attractively fresh and tasty, while a dry and soft apple may feel insipid. Such tactile sensations and preferences are derived from touch- and pressure-sensitive mechanoreceptors located within the oral mucosa and periodontium and from stretch and other receptors in the masticatory muscles, temporomandibular joints and periosteum.

Oral mechanoreceptors also contribute to discrimination among food materials. Tactile and kinaesthetic sensations are essential for accurate identification of the size, shape, hardness, softness and other textures of solids and liquids in the mouth. This information may be useful to determine whether these materials are beneficial or harmful to ingest. In addition, tactile oral sensations are indispensable for the control of movements of the jaw, tongue and hyoid during chewing and swallowing, as well as for salivary secretion.

The lips and tongue are most sensitive areas in the body, and as sensitive as the fingertips. The teeth have no sensory receptors on their surfaces, but there are sensitive receptors within periodontal ligaments. Besides these sensory receptors in the oral mucosa, we shall also consider receptors in the extraoral tissues including masticatory musculature and temporomandibular joints.

6.3. Mechanosensitive receptors in oral structures

6.3.1. Sensory receptors in the oral mucosa

(a) Morphology, distribution and innervation

Mechanoreceptors in the skin include free nerve endings and complex receptors. The free nerve endings act as receptors for pain and temperature, but partly serve also for mechanoreception. Complex structures, including Merkel's disc, tactile hairs, Pacinian corpuscle, Meissner's corpuscle, Krause's corpuscle, and Ruffini end-organs, increase sensitivities to touch, pressure, and vibration (Grossman and Hattis, 1967). These structured endings also exist deep in the muscles, tendons and joints, and sense deformation of tissues. Muscles and tendons also have muscle spindles and Golgi tendon organs to detect change in length and tension of muscles. All kinds of mechanoreceptor except Pacinian corpuscles and tactile hair are located in the oral mucosa. Most of them are Meissner's corpuscles or Krause's corpuscles. Morphologically, similar nerve endings are observed in gingival tissues, tongue and palate (Gairns, 1955; Grossman and Hattis, 1967).

Mechanoreceptors are high in density in the front of the oral cavity and are sparse in the posterior part. The tip of the tongue shows the highest density of receptors in the oral mucosa. On the dorsum of the tongue and on the palate, the mechanoreceptors are sparse in the middle part but abundant in the posterior part, such as the anterior surface of the uvula, probably because of their role in swallowing.

Tactile information from the facial, oral, and tongue regions is transmitted to the brain via the trigeminal nerve. Receptors in the mucous membrane of the epiglottis and the isthmus are innervated by the glossopharyngeal nerve.

(b) Response properties of oral mechanoreceptors

Sensory receptors are physiologically classified into two categories: fast adapting receptors which respond only at the moment when the stimulus is applied and slow-adapting receptors which continue to respond as long as the stimulus is applied.

The encapsulated endings such as Meissner's and Krause's corpuscles are fast adapting receptors which detect the velocity and acceleration of touch stimuli. In contrast, Merkel cell–neurite complexes are slowly adapting pressure receptors and serve to detect the velocity of displacement. Free nerve endings have fast and slowly adapting types. The amount of deformation of the mucous membrane and the impulse frequency generated in the sensory neurones can be related according to psychophysical principles (Sakada, 1983).

How do oral mechanoreceptors act then during the chewing of foods? Appenteng and his associates (1982) found that the receptive fields of mechanoreceptors located in the tongue and oral mucosa of rabbits ranged between 1 mm^2 or smaller and 30 mm^2. Thresholds for stimulation had a range of about 1 g to 20 g, with the majority less than 3 g. Mechanical contact with the food bolus during chewing gives an excitatory effect in these receptors. Accordingly, these mechanoreceptors commence their activity at the closing phase of a chewing stroke, and it is sustained during the intercuspal phase and ends in the opening phase. These receptors are not excited even when the jaw moves during the remainder of the chewing rhythm. Hence, it seems that the role of these receptors is not to detect jaw position or velocity but to sense the position of the food bolus in the oral cavity.

6.3.2 Periodontal mechanoreceptors

(a) Anatomy and innervation

The texture of food can be perceived during chewing by excitation of mechanoreceptors in the periodontal ligaments. In man, these include free nerve endings and four types of special endings, Ruffini corpuscles, coil endings, spindle endings and end bulk endings (Maeda, 1990). When a tooth is slightly displaced during chewing, the periodontal ligaments are stretched, deforming and exciting the receptors.

Sensory neurons from this type of mechanoreceptor ascend to the brain stem via the trigeminal nerve. There are two afferent pathways. One passes to the trigeminal motoneurones via the trigeminal mesencephalic nucleus, as does the jaw muscle sense, while the other passes to the trigeminal motoneurones via the trigeminal brain-stem sensory nuclei. These periodontal afferents may be monosynaptic or polysynaptic. Also, the sensory neurones for these two types of afferents have been confirmed to come from the respective receptors (Byers, 1989).

(b) Physiological response

The response properties of the periodontal afferents have been investigated in several species of mammals, including human beings (Daunton, 1977; Hannam, 1976; Johansson and Olsson, 1976).

A fast adapting type responds only at the moment of when the tooth is pushed; the slowly adapting type responds while the tooth is pushed, and a variant type of the latter has been verified (Fig. 6.1A). In rabbits, slow-adapting receptors

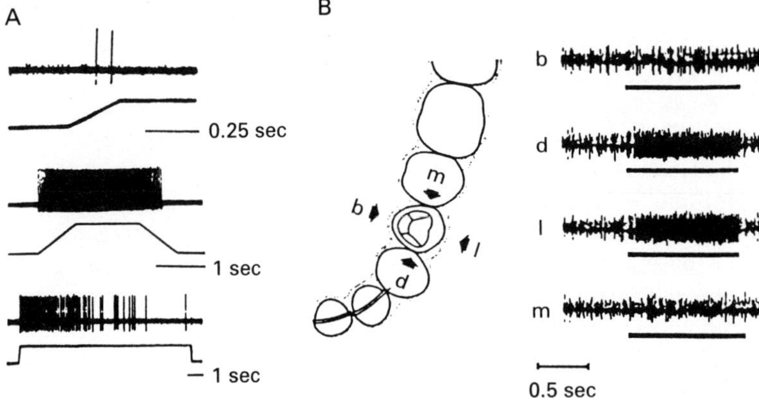

FIG. 6.1.   Response of periodontal receptors. A. Three types of receptors with different adaptabilities to tooth press. *Top*: fast-adapting, *middle*: very slowly adapting, *bottom*: slowly adapting receptor. (From Hannam, 1976.) B. Difference in response of human periodontal receptor according to the direction of tooth press. A force of around 2N was applied in the buccal (b), distal (d), lingual (l) and mesial (m) directions. This receptor responded particularly well to press applied in the distal and lingual directions. (From Johansson and Olsson, 1976.)

have been reported to be abundant in the anterior segment and fast adaptive receptors in the posterior segment.

A particular periodontal afferent responds maximally to one particular direction of applied force. (Fig. 6.1B). There are also differences in response between fibres with respect to the magnitude of stimuli applied but the discharge frequency of a periodontal afferent increases in proportion to the strength of stimulus within a limited range (Appenteng *et al.*, 1982). The threshold for a pressure stimulus applied to a tooth is estimated to be greater than 1 g. The perceptual threshold level is lowest for incisors or canines, and high for the posterior teeth (Loewenstein and Rathkamp, 1955; Bonagro *et al.*, 1969), but this level is influenced by the velocity of pressure application (van Steenberghe and de Vries, 1978).

Appenteng *et al.* (1982) reported, in rabbits, that no signals are transmitted from the receptors while the upper and lower teeth are separated, but firing rate increases in accordance with a biting force once the teeth are in contact.

6.4. Sensory receptors in masticatory muscles and temporomandibular joints

6.4.1. Muscle spindle

The muscle spindle is a receptor located in skeletal muscles, excited by the stretching of its sensory endings. It may be several millimetres in length (those located in the masticatory muscles are slightly shorter) and rarely over 100 µm

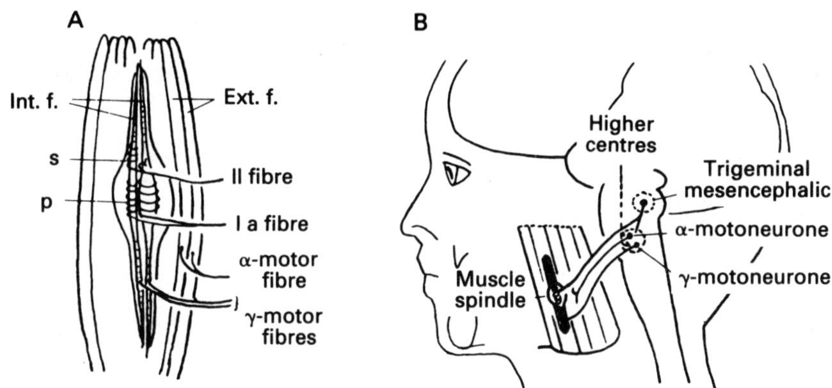

FIG. 6.2.   Structure of muscle spindle and the jaw-jerk reflex arc with gamma innervation. A. Schematic representation of the muscle spindle. Intrafusal muscle fibres are innervated by thin gamma motor fibres, while extrafusal ones are by thick alpha motor fibres. Both primary and secondary endings are located at or near the centre of the spindle where cell nuclei are gathered. P, primary ending; S, secondary ending; Int. f., intrafusal muscle fibres; Ext. f., extrafusal muscle fibres. B. Jaw-jerk reflex arc is composed of the muscle spindle in the jaw-closing muscles, Ia or II afferent fibres whose cell bodies are located in the trigeminal mesencephalic nucleus, the trigeminal jaw-closer motoneurones, the trigeminal motor nerve and jaw-closing muscles. This reflex operates either when the jaw opens or when the gamma motoneurones are excited.

wide and is surrounded by a spindle-like capsule of connective tissue. The histology of this receptor is shown schematically in Fig. 6.2A. Muscle spindles lie in parallel with the ordinary (extrafusal) muscle fibres. Inside the capsule, a few thinner muscle fibres (intrafusal fibres) run along the long axis. The spindle's centre contains large groups of nuclei. Because very few myofibrils, however, are present in this region, there is little contractility.

Termination of the sensory afferents has two modes. The annulospiral primary ending ends in the central region of the intrafusal fibres. The spray-like secondary sensory ending is located at some distance from the centre.

Physiological properties of muscle spindles in the jaw-muscles were recently reviewed by Taylor (1990). While alpha efferents act as motor supply to extrafusal muscles, the intrafusal muscles are supplied by the smaller gamma efferents. If the entire muscle is stretched passively or sensory nerve endings are stretched by the contraction of intrafusal muscle fibres, action potentials are generated in the spindle receptors. The excitation of the nerve endings is transmitted to the central nervous system via Ia fibres from the primary endings and II fibres from the secondary endings. Because of the difference in responses, the primary endings are considered to transmit information regarding velocity and muscle length, while the secondary endings sense muscle length. Since muscle spindles lie in parallel with the extrafusal muscle fibres, they are liable to relax as extrafusal muscles contract. However, the responsiveness of the sensory

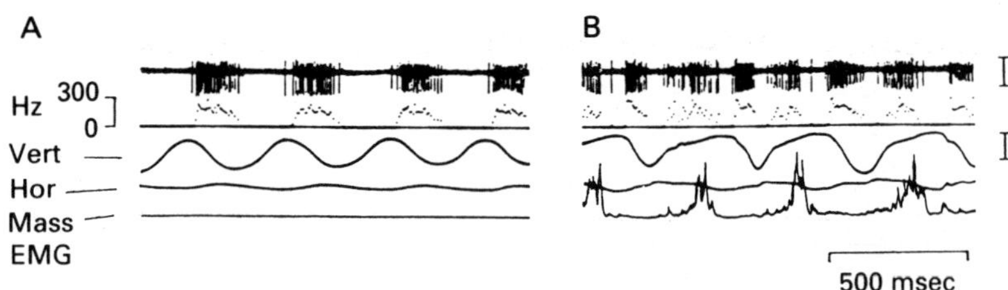

FIG. 6.3. Muscle spindle responses recorded in the mesencephalic nucleus of the monkey. A. Response during depression of the mandible by an experimenter. B. The spindle responds during the jaw closing phase as well as during the jaw opening phase in a masticatory cycle of biscuit chewing. (From Goodwin and Luschei, 1975.)

endings does not decrease; rather, it may increase because gamma-motoneurones are frequently activated in concert with alpha-motoneurones (alpha-gamma co-activation) during voluntary movement and muscular reflexes (Fig. 6.2B).

Muscle spindles in the masticatory musculatures anatomically differ from those in the limb muscles at least two ways. First, most of the spindles are located in jaw-closing muscles (such as masseter, temporalis, medial pterygoid, and partly lateral pterygoid muscles) and very few in jaw-opening muscles (Kubota and Masegi, 1977). Secondly, some of the sensory nerve (Ia and II) fibres from the muscle spindles in jaw-closing muscles have their cell bodies in the brain and form the trigeminal mesencephalic nucleus. This contrasts with spindles in muscles of the limbs where the sensory cell bodies are in spinal ganglia.

Physiological responses of primary and secondary endings of the muscle spindles in the jaw-closing musculature, recorded from the trigeminal mecencephalic nucleus, are similar to those of the spindles in limb muscles. The primary ending shows a transient dynamic response followed by a static response when the ending is stretched during the act of jaw-opening. In contrast, the secondary ending provides primarily a static response, with almost no dynamic response. There are also a lot of spindle endings in the jaw-closing muscles that show properties intermediate to those two responses during jaw-opening (Cody, Lee and Taylor, 1972).

Sensitivity of the spindles endings is controlled by activity of the gamma motor fibres that innervate the intrafusal muscle fibres. Consequently, the response of spindle endings during passive stretching differs from those during active stretching caused by gamma activation. Figure 6.3 represents sensory signals of a muscle spindle recorded from the trigeminal mesencephalic nucleus in the monkey. As shown in Fig. 6.3A, the muscle spindle transmits signals only during the jaw-opening phase in case of passive open/close jaw movements. However, as shown in Fig. 6.3B, signals are also conveyed during the jaw-closing and intercuspal phases when the monkey performed chewing efforts on a biscuit. The responsiveness of a muscle spindle during voluntary contraction

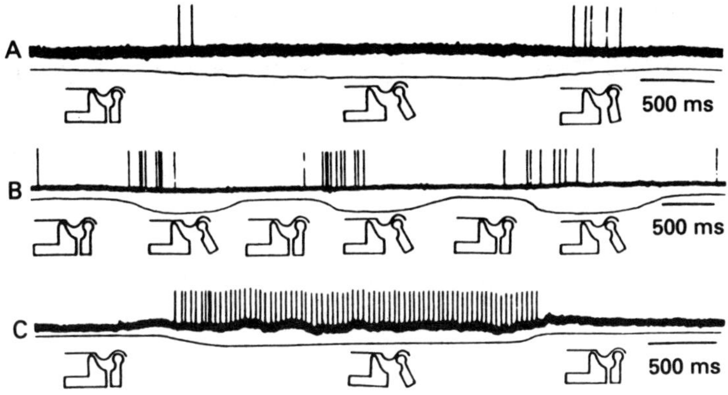

FIG. 6.4.   Response of the temporo-mandibular joint receptors of the cat. A. Fast adapting receptor (on-off type). B. Fast adapting receptor (on type). C. Slow adapting receptor. Opening, rest or closing of the joint is represented by the direction of the movement of the isolated condyle. (From Kawamura and Abe, 1974.)

of the muscles is obviously intensified by gamma activation which occurs simultaneously with activation of alpha motoneurones innervating the jaw-closing muscles. Such muscle spindle activity during jaw-closing and intercuspal phases is considered to increase jaw-closing muscle activity during clenching of teeth, via a jaw-jerk reflex arc (cf. the knee-jerk test).

There is a proportional relationship between the amount of jaw-opening and the magnitude of static response of the spindle endings (Inoue *et al.*, 1981). This indicates a contribution of muscle spindles to the perception of jaw position. Study of the human mandibular position sense has supported this interpretation (Morimoto, 1983).

6.4.2. *Golgi tendon organ*

The Golgi tendon organ is a tree-like sensory ending enclosed in a spindle-like connective tissue capsule, that lies near the junction of a tendon with a muscle. In man, some 10 to 20 muscle fibres are connected to one tendon organ. A typical tendon organ in limb muscles has an ending of about 0.5 mm in length. The endings are supplied by Ib afferent fibres (thinner than Ia) of high conduction velocity.

In contrast to muscle spindles, the Golgi tendon organ lies in series with skeletal muscle fibres and therefore discharges during passive stretching of the muscle as well as when the tendon is stretched by the contraction of the muscle. Accordingly, the tendon organ is considered to be a muscle tension receptor rather than a muscle stretch receptor. The response threshold level of the tendon organ for the contraction of muscle is very much lower (only 4 mg) than that for the passive stretching of the muscle (Jansen and Rudjord, 1964; Binder *et al.*,

1977). This indicates that the tendon organ plays a role in transmitting the magnitude of contractile force during movement.

The tendon organ also provides a highly sensitive response to slow contraction of the muscle. The organ increases its activity in proportion to the tension applied, with the ratio of increase gain differing among tendon organs. In addition, the number of tendon organs recruited increases as the muscle tension increases (Crago *et al.*, 1982).

Few investigations have been reported on the Golgi tendon organs in the masticatory muscles. Morphological studies have verified the existence of this organ in the masseter and temporalis muscles in cats (Kawamura and Hamada, 1974; Lund *et al.*, 1978). The tendon organ shows localization similar to the muscle spindle and lies deep in the muscle (Lund and Matthews, 1981). It is usually connected with a muscle spindle but the significance of such a linkage is not understood. It is probable that gamma-efferents that influence muscle spindles can also influence the Golgi tendon organ as well.

The physiological role of Golgi tendon organs in masticatory muscle activity remains unknown. However, it may contribute to the control of chewing force.

6.4.3. *Receptors in the temporomandibular joint*

Temporomandibular joint receptors are mostly located in the joint capsule and ligaments. Morphologically, four types of receptors (free nerve ending, Ruffini ending, Golgi tendon organ, and Pacinian corpuscle) have been verified (Thilander, 1961; Klineberg *et al.*, 1970).

The Ruffini ending, or the Type I receptor, may transmit information regarding the position of the joint.

The Pacinian corpuscle, or the Type II receptor, located in the temporomandibular joint, is considerably smaller (180×20 μm on average) than its subcutaneous homologue which acts as a touch/vibration receptor. Because it shows rapidly adapting discharge at a low threshold level of pressure on the joint capsule, the Pacinian corpuscle appears to convey signals regarding joint movement.

The Golgi tendon organ, or the Type III receptor, is similar to the one that exists in the ligaments of the muscle. It is innervated by the cutaneous nerve and this responds as the ligament is stretched. Therefore, this organ is assumed to transmit information regarding joint position (angle).

Free nerve endings, or the Type IV receptors, are branches of unmyelinated Group IV nerve fibres (smaller than 1 μm in diameter) that end in the joint and have a broad range of distribution from each fibre. These receptors have a high threshold to mechanical stimuli and respond to materials such as bradykinin and serotonin injected into the artery that selectively stimulate noxious receptors. Because of this, free nerve endings are considered to be receptors for noxious stimuli and to relate to pain sensation in the joint. In an electrophysiological experiment (Kawamura and Abe, 1974), responses from a joint receptor were recorded through the auriculo-temporal nerve, and both the slow-adapting and fast-adapting types of receptor were identified (Fig. 6.4), but there were extremely few of the latter. The fast-adapting type responded only at the onset and/or cessation of jaw movement, regardless of the direction of movement. The slow-adapting type responded to the rotation or medial displacement of the temporomandibular joint; there was a jaw position at which the maximum response was recorded and the magnitude of response was proportional to the rotational velocity and angle of the joint. Hence, the slowly adapting type is considered to transmit information regarding the rotational velocity of the condyle and the angle of jaw-opening to the central nervous system.

6.5. Tactile control of jaw and tongue movements

6.5.1. *Effects of disrupted oral sensory input*

If oral sensory information contributes to the control of mastication, disrupting the input should change chewing function. In man, it becomes difficult to place and keep a food bolus on the occlusal platform after local anaesthetization of gingiva, periodontal ligament and temporomandibular joint (Schaerer, 1966).

In pigeons and rats, severing the trigeminal sensory branches (particularly the maxillary and mandibular nerves) produces insufficient jaw-opening movement and protrusion of the tongue (Zeigler, 1975; Miller, 1981; Jacquin and Zeigler, 1983), and hence food uptake into the oral cavity becomes difficult. In addition, oral sensory denervation makes it difficult to comminute and grind food taken into the mouth (Inoue *et al.*, 1989). Figure 6.5A represents a jaw movement pattern in the coronal plane. Both the trajectories in the vertical and lateral directions become smaller, more irregular and more complicated compared with those recorded before severing the nerve. Another finding which should be noted is that, as shown in the record of a jaw movement in the vertical direction in Fig. 6.5B (top), the rabbit could occlude the upper and lower teeth firmly in every chewing stroke before severing the nerve; however, after severing the nerve, the rabbit could not occlude the teeth firmly, stopped closing the jaw before it was shut and began to open it (Fig. 6.5B, bottom). This indicates that the chewing force had become weakened. Consequently, chewing efficiency decreased and the number of chewing strokes before swallowing of food (i.e. the swallowing threshold) increased. Electromyographic records showed a remarkable decrease in masseter muscle activity after severing the nerve, greatest one week after surgery.

On the other hand, denervation of facial skin (including the lips) by severing the infraorbital and mental nerves did not change the jaw displacement pattern or masticatory muscle activity in rabbits.

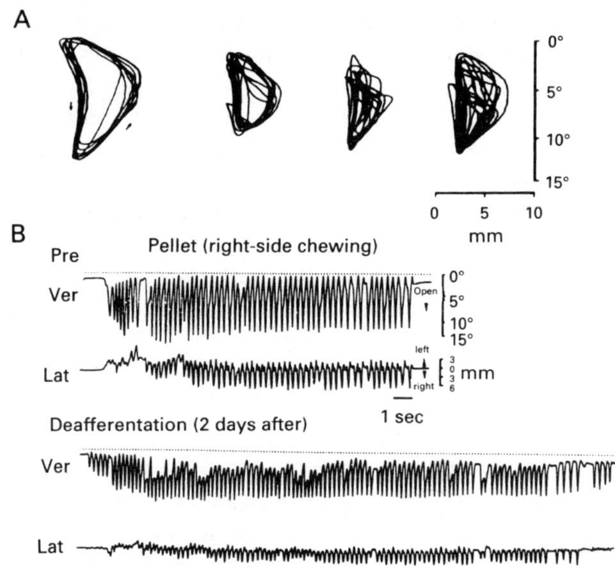

Fig. 6.5. Effects of trigeminal deafferentation on the chewing movements of the mandible. A. Jaw movements in the frontal plane. The traces from the left are control, two days after lesion and one week and two weeks after lesion. B. Vertical and horizontal components of the jaw movements in a masticatory sequence. *Upper traces*: before deafferentation; *lower traces*: after deafferentation. (From Inoue *et al.*, 1989.)

6.5.2. Effects of stimulation of oral sensory input

(a) Effects of oral stimulation on rhythmic jaw movement

Electrical stimulation of the cortical masticatory area in anaesthetized rabbits induces rhythmic jaw movement similar to that observed during chewing. If an elastic material is inserted between the upper and lower posterior teeth, the jaw movement patterns and jaw-closing muscle activity are altered, as shown in Fig. 6.6. Figure 6.6A1 and 6.6A2 illustrate jaw movement patterns in the frontal (coronal) view before and during chewing the material, respectively. Figure 6.6A3 superimposes these two patterns. During chewing of the material, the horizontal component, i.e. the width of grinding jaw movement has increased. Activity in the masseter muscle (jaw-closing muscle) increased during chewing of the test material relative to activity before the insertion of the material, while significant change was not visible in digastric muscle (jaw-opening muscle) activity (Fig. 6.6). In addition, the jaw-closing muscle activity increased in proportion to the thickness and hardness of the test material.

In this experiment, the change in jaw-movement patterns and increase in jaw-closing muscle activity during chewing of the test material were involuntary because the animal was anaesthetized. The insertion of the material

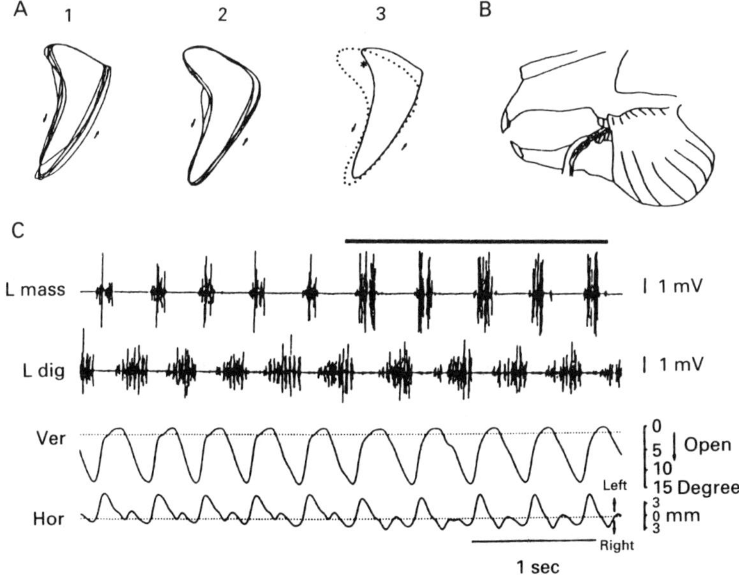

FIG. 6.6.   Modification of the pattern of cortically induced rhythmic jaw movements and associated EMG activities of the masseter and digastric muscles during application of the test strip between opposing molars. A. Patterns of jaw movements before (1) and during (2) application of the test strip and their superimposed tracings (3). B. The procedure for application of a polyurethane-foam strip between the molars using a forceps. C. EMG activities of the masseter and digastric muscles recorded simultaneously with vertical and horizontal jaw movements. The thick bar indicates the period of test strip application. L Mass, left masseter muscle; L Dig, left digastric muscle; Ver and Hor, vertical and horizontal jaw movements. (From Morimoto *et al.*, 1989.)

stimulated the periodontal receptors. Therefore, the oral sensory input automatically contributes to the control of masticatory force according to the nature of the food. Also, this control system has a positive feedback mechanism which enables increase in masticatory force as the food material is crushed.

Such an influence of oral stimulation on the masticatory apparatus decreases after anaesthetizing the trigeminal sensory branch. Nevertheless, people who have lost their teeth can control masticatory force. This indicates that there are contributions besides tactile sensation to the masticatory force control system. Muscle spindles in the jaw-closing muscles may participate. Sensory denervation of muscle spindles by lesions in the trigeminal mesencephalic nucleus produces decreases in jaw-closing muscle activity and masticatory force. In addition, in an animal whose intraoral sensation is disrupted by severing the trigeminal sensory branch, trigeminal mesencephalic nucleus lesions disturb the control of the masticatory force during chewing of test materials (Morimoto *et al.*, 1989). Consequently, it can be concluded that the masticatory force is

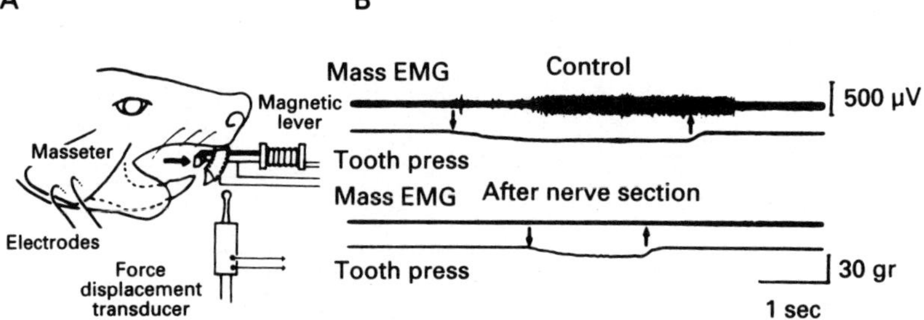

FIG. 6.7.   Tonic periodontal masseteric reflex induced by pressing ipsilateral maxillary incisor. A. Methods of force application and EMG recording. B. responses before and after nerve dissection. *Upper trace*: the response of masseter muscle before nerve section. *Lower trace*: disappearance of the response after dissection of the maxillary nerve. (From Funakoshi and Amano, 1974.)

controlled by both periodontal receptors and muscle spindles in the jaw-closing musculature.

(b) Jaw reflexes

Periodontal masseteric reflex. When the labial surface of the upper incisor is lightly tapped or the associated gingival tissue is stimulated electrically during light clenching of teeth, a reflexive electric potential is evoked in the masseter muscle. This is termed the periodontal masseteric reflex. The receptors are in the periodontal ligament (Goldberg, 1971). This reflex is found not only in man but also in rats, cats and monkeys.

In rats, continuous activity with long latency together with a transitory excitation is evoked in the masseter muscle when the upper incisor is pressed continuously in a linguo-labial direction (Fig. 6.7). This reflex is called as the tonic periodontal-jaw muscle reflex (Funakoshi and Amano, 1974). This response is influenced by the direction of a force exerted to the teeth. That is, if the upper incisor is pushed labio-lingually, the jaw-closing muscle activity is inhibited.

The periodontal-masseteric reflex arc may be monosynaptic via the trigeminal mesencephalic nucleus. The pathway for the tonic periodontal-jaw muscle reflex involves the periodontal ligament, the trigeminal ganglion, the trigeminal brain stem sensory complex, and jaw-closing motoneurones. These reflexes possibly contribute to the control of chewing force.

Jaw-jerk reflex. A jerk of the jaw is initiated by lightly tapping a small piece of board placed between the upper and lower teeth or against the chin. The transient stretching of jaw-closing musculatures produces a contraction of the jaw-closing muscle. This is a stretch reflex similar to the knee-jerk reflex. The muscle spindle is the sensory receptor and the reflex pathway goes via the trigeminal mesencephalic nucleus to jaw-closing motoneurones and the jaw-closing musculature (Fig. 6.2B). The reflex is usually monosynaptic, but a polysynaptic pathway also exists. The jaw-jerk reflex differs from other stretch reflexes in lacking reciprocal inhibition on the jaw-opening motoneurones. Moreover, a stretch reflex is not initiated by stretching the jaw-opening musculatures, since they contain very few spindles.

Various suggestions have been made regarding the physiological significance of the connections producing the jaw-jerk reflex. They may control the contraction force of jaw-closing musculatures promptly in response to the loading to the lower jaw produced by a bite during chewing of food or the unloading that is generated at the moment of crushing food (Lamarre and Lund, 1975). The loading reflex is important to control the chewing force in response to the hardness and consistency of food (Lavigne *et al.*, 1987). With an unloading reflex, on the other hand, an immediate reduction of force exerted on the lower jaw would arise at the moment of crushing hard bolus from a decrease in facilitation of jaw-closing musculatures via the jaw-jerk reflex, because muscle spindles have been relaxed by the sudden shortening of jaw-closing musculatures. A possible second function of the jaw-jerk reflex is to maintain jaw posture, i.e. the mandibular resting position. When the mandible is destabilized during running and jumping activities, the muscle spindle is activated and thus stiffens jaw-closing musculatures and resists instability (Goodwin *et al.*, 1978).

6.6. Physiology of tongue movement and sensation

The tongue is the most tactually sensitive organ in the body. In four-footed animals and human infants who cannot use their fingers efficiently, the tip of the tongue is used as an alternative to the finger tip.

Moreover, the tongue is a sophisticated motor organ. It not only moves rhythmically in concert with the lower jaw but also places the food bolus on the occlusal platform of molar teeth on the chewing side and transports it from one side to the other. Furthermore, one seldom bites one's own tongue.

We are not usually conscious of such intricate movements of the tongue and their sensory control. However, inflammation of the tongue (stomatitis) makes it feel very difficult to chew and to speak.

6.6.1. Anatomy and physiology of tongue musculatures

The tongue can perform complicated movements because of its unique constitution of muscles. Extrinsic muscles attach to the cranium, the mandible or the hyoid bone on one side and to the mucous membrane of the tongue on the other (Fig. 6.8A), and thus can change the overall tongue position. Intrinsic muscles terminate in the mucosa or in other muscles of the tongue (Fig. 6.8B). By virtue of these structural characteristics, the tongue can alter its shape in a delicate way.

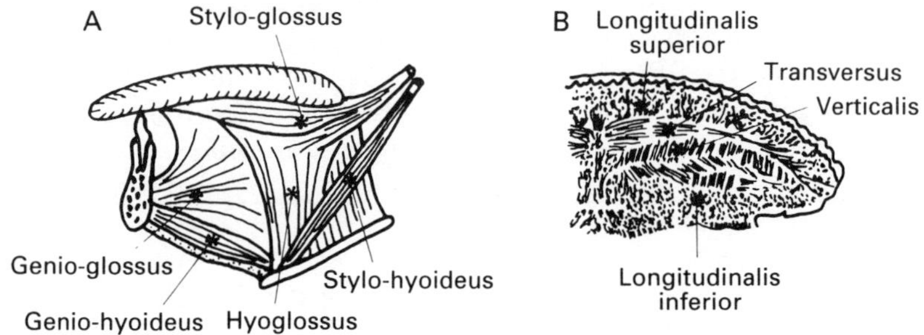

FIG. 6.8.   Tongue muscle constructions. A. Extrinsic tongue muscles. B. Intrinsic tongue muscles.

The physiology of the tongue's musculature serves rapid movements rather than tonic maintenance of tongue position.

6.6.2. *Tongue movement during chewing*

Tongue movement during the act of chewing in man was divided into five phases on the basis of observations of a subject who had lost teeth on one side (Abd-El-Malek, 1955):

Phase 1 (preparatory stage, Fig. 6.9A). The tongue alters its shape from resting posture into a trough-like structure and collects the food bolus on its dorsal surface.

Phase 2 (throwing stage, Fig. 6.9B). The anterior portion of the tongue with the food bolus on its dorsal surface is twisted toward the chewing side and the dorsum contacts the lateral surface of the teeth. In this way, the food bolus is placed on the occlusal platform of the posterior teeth.

Phase 3 (guarding stage). The tongue presses its dorsum against the medial surface of the teeth, while twisting to prevent the food from slipping off from the occlusal surface (Fig. 6.9C).

Phase 4 (sorting-out stage). This begins when the upper and lower teeth are separated at the end of the intercuspal phase. Since the buccinator muscle retracts the cheek toward the inside, the buccal mucosa is held between the upper and lower teeth and presses the food bolus against the buccal cavity and the tongue. In this phase, the tongue moves again rapidly to select a large food bolus which requires further chewing and the fractions of food are placed again on the occlusal surface. Those fully ground are collected onto the lateral margin of the tongue. This movement is repeated until the food is completely ground.

Phase 5 (bolus formation stage). The phases 4 and 5 partly overlap. The tongue mixes saliva with the crushed food by alternations from one side to the other, and coats the bolus with mucus. Then, the tongue assumes the posture to be followed by a swallowing action.

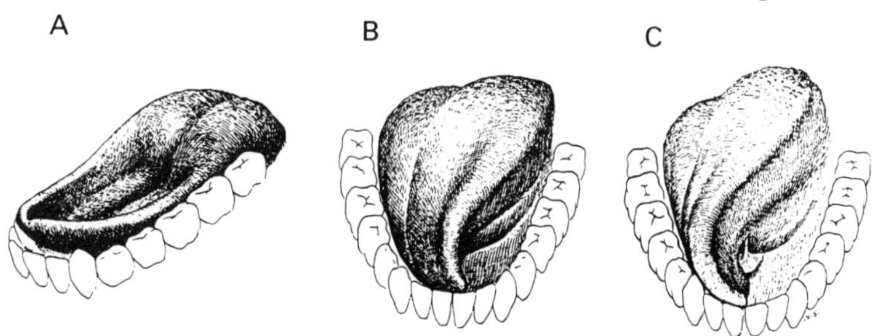

F IG. 6.9.   Pattern of tongue movements in a masticatory sequence. A. Preparatory stage.
B. Throwing stage. C. Guarding stage. (From Abd-El-Malek, 1955.)

6.6.3. Sensation

The tongue is practically as sensitive to touch as the finger tip, which is the
tactually most sensitive area in the body. The threshold of two-point discrimination is often less than 1 mm (Sherrington, 1900). On the dorsal surface of the
tongue, there are numerous papillae containing structured sensory endings.
Also, free nerve endings are frequently found. Although the functions of the
different histological types are not fully understood, it appears that the complex
endings transmit the senses of touch and pressure, and the free endings convey
temperature and pain.

Besides these sensory receptors, there are proprioceptors in the tongue
muscle. Abundant muscle spindles have been reported, particularly in the
human tongue (Kubota *et al.*, 1975). Other mammals, however, have very few
lingual spindles, and so perhaps spindles in the human tongue muscles
contribute to intricate movements such as in speech. Nevertheless it is also
perhaps conceivable that they could contribute to the perception of food texture.

6.6.6. Jaw–tongue reflex

When the mouth of an anaesthetized animal is opened passively, the tongue is
retracted posteriorly and the root of the tongue is elevated. This reflex has been
termed as the jaw–tongue reflex (Schoen, 1931). It is evoked by excitation of
sensory receptors in the temporal muscle (Morimoto *et al.*, 1978).

In cats, when rhythmic jaw and tongue movements are elicited by electrical
stimulation of masticatory area in the cerebral cortex, the tongue is retracted in
concert with rapid jaw opening.

Accordingly, the position and movement of the tongue are controlled
centrally and peripherally so that it is not bitten.

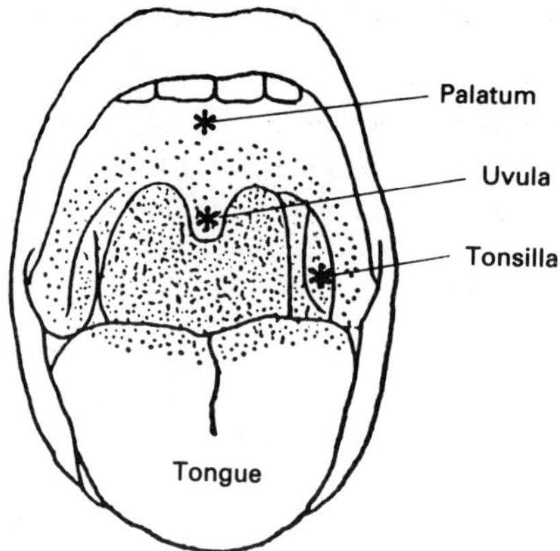

FIG. 6.10.   The areas which induce swallowing. Swallowing is easily induced by mechanical stimulation to the densely dotted area.

6.7. Swallowing and oral mechanoreception

The swallowing process can be divided into three phases, oral, pharyngeal, and oesophageal, according to the position of food bolus. In the oral phase, the food bolus is formed between the tongue and the hard palate; this process is under voluntary control. Once the food bolus reaches the epiglottis, the pharyngeal phase begins. The process thereafter is initiated reflexively. Mechanoreceptors in the pharyngeal mucosa must be stimulated to elicit the swallowing reflex (Fig. 6.10). Afferents project to the reticular formation on each side of the brain-stem near the inferior olivary nucleus, and a programme of muscle activity is evoked in the following sequence. (1) The posterior tongue is elevated to cut off the pharynx from the oral cavity. (2) The soft palate is raised to obstruct the nasopharyngeal cavity and thus to divide the nasal cavity from the pharynx. (3) The pharyngeal opening of auditory tube is closed. (4) The pharynx is elevated anterosuperiorly, and the epiglottis is rotated downward to cover the laryngeal opening, thus excluding the pharyngeal and laryngeal cavities; the glottis is also closed. (5) A negative pressure is produced in the superior part of the oesophagus by the relaxation of pharyngeal constrictors, and the food bolus is propelled into the oesophagus. The process of swallowing is completed in 0.3–0.5 second.

In man, swallowing occurs about 600 times per day, including during sleep. When the activity in the brain stem reticular formation and/or in the cortex is altered because of general anaesthesia, drunkenness, or sleep, the swallowing centre is weakened or its function even disappears, and disorganized swallowing movements occur. Local anaesthesia affecting sensory input from the oral and the pharyngeal regions results in disordered swallowing.

The swallowing reflex matures and so swallowing disorder may occur in immature or new-born infants.

6.8. Oral mechanoreceptors and preference

Each person has to some degree his or her own preferences among foods. Also, each culture has foods of distinctive taste, smell and texture. Be a food ever so favoured by one people, it is not necessarily favoured at all and may even be disliked by peoples who are unfamiliar with that sort of food. A good example is the texture of raw or marinated fish which is popular in Japan but not necessarily loved by Westerners.

Such idiosyncratic preferences for foods appear to be explained by learning to recognize patterns of sensory stimulation from taste and texture, etc. Exposure to tastes, smells and textures of foods in infancy induces preferences but these often alter as one grows up and has wider experiences. Some learn to like a food disliked in childhood when other people are frequently seen to enjoy eating it. Furthermore, a favourite food may become distasteful if one has too much of it or it is associated with a bad experience.

Physical properties of food such as those detected by pressure on the teeth and the tongue are not necessarily preferred or disliked in themselves, as are sweetness and bitterness. For example, the crispness that would be considered palatable in an apple would be unattractive in a strawberry. Also, preferences differ among foods which have similar physical properties because of the numerous other factors that differ, such as colour, aroma, taste and temperature and the past experiences of eating different flavours of the same texture. The learned specificity of preference for a food means that relation between oral mechanoreception and preference is not simple. Oral mechanoreception and motor control mediated in the brainstem are probably modulated by higher centres in the recognition of and the preferences for particular foods by different individuals. We do not yet have the proper experimental methods to elucidate this neurophysiologically.

References

Abd-El-Malek, S. (1955). The part played by the tongue in mastication and deglutition. *Journal of Anatomy*, **89**, 250–254.

Appenteng, K., Lund, J. P. and Seguin, J. J. (1982). Intraoral mechanoreceptor activity during jaw movement in the anesthetized rabbit. *Journal of Neurophysiology*, **48**, 27–37.

Binder, M. D., Kroin, J. S., Moore, G. P. and Stuart, D. G. (1977). The response of Golgi tendon organs to single motor unit contractions. *Journal of Physiology*, **271**, 337–349.

Bonagro, J. G., Dusza, G. R. and Bowman, D. C. (1969). Ability of human subjects to discriminate forces applied to certain teeth, *Journal of Dental Research*, **48**, 236–241.

Byers, M. R. and Dong, W. K. (1989). Comparison of trigeminal receptor location and structure in the periodontal ligament of different types of teeth from the rat, cat and monkey, *Journal of Comparative Neurology*, **279**, 117–127.

Cody, F.W.J., Lee, R.W.H. and Taylor, A. (1972). A functional analysis of the components of the fifth nerve in the cat. *Journal of Physiology* 226, 249–261.

Crago, P. E., Houk, J. and Rymer, W.Z. (1982). Sampling of total muscle force by tendon organs, *Journal of Neurophysiology*, **47**, 1069–1083.

Daunton, N.G. (1977). Sensory components of bite-force response in the rat. *Journal of Comparative Psychology*, **91**, 203–220.

Funakoshi, M. and Amano, N. (1974). Periodontal jaw muscle reflexes in the albino rat. *Journal of Dental Research*, **53**, 598–605.

Gairns, F.W. (1955). The sensory nerve endings of the human palate. *Quarterly Journal Experimental Physiology*, **40**, 40–48.

Goldberg, L.J. (1971). Masseter muscle excitation induced by stimulation of periodontal and gingival receptors in man, *Brain Research*, **32**, 369–381.

Goodwin, G.M. and Luschei, E.S. (1975). Discharge of spindle afferents from jaw closing muscles during chewing in alert monkeys, *Journal of Neurophysiology*, **38**, 560–571.

Grossman, R.C. and Hattis, B.F. (1967). Oral mucosal sensory innorvation and sensory experience. In J. F. Bosma (Ed). *Oral Sensation and Perception* (pp. 5–62), Charles C. Thomas: Illinois.

Hannam, A. G. (1976). Periodontal mechanoreceptors. In D. J. Anderson and B. Matthews (Eds), *Mastication* (pp 42–49) Wright: Bristol.

Inoue, H., Morimoto, T. and Kawamura, Y. (1981). Response characteristics and classification of muscle spindles of the masseter muscle in the cat. *Experimental Neurology*, **74**, 548–569.

Inoue, T., Kato, T., Masuda, T., Nagashima, T., Kawamura, Y. and Morimoto, T. (1989). Modifications of masticatory behavior after trigeminal differentiation in the rabbit, *Experimental Brain Research*, **74**, 579–591.

Jacquin, M. F. and Zeigler, H. P. (1983). Trigeminal orosensation and ingestive behavior in the rat, *Behavioral Neuroscience*, **97**, 62–97.

Jansen, J.K.S. and Rudjord, T. (1964). On the silent period and Golgi tendon organs of the soleus muscle of the cat, *Acta Physiologica Scandinavica*, **62**, 364–379.

Johansson, R.S. and Olsson, K.A. (1976). Microelectrode recordings from human oral mechanoreceptors, *Brain Research*, **118**, 307–311.

Kawamura, Y. and Abe, K. (1974). Role of sensory information from temporomandibular joint. *Bulletin of Tokyo Medical Dental University*, **21**, (Supplement), 78–82.

Kawamura, Y. and Hamada, T. (1974). Studies on morphology and functions of Golgi tendon organ of masticating muscle in cat, *Bulletin of Tokyo Medical and Dental University*, **21**(Supplement), 7–10.

Klineberg, I. J., Greenfield, B. E. and Wyke, B. D. (1970). Contribution to the reflex control of mastication from mechanoreceptors in the temporomandibular joint capsule. *Dental Practitioner*, **21**, 73–83.

Kubota, K., Negishi, T. and Masegi, T. (1975). Topographical distribution of muscle spindles in the human tongue and its significance in proprioception. *Bulletin of Tokyo Medical and Dental University*, **22**, 235–242.

Kubota, K. and Masegi, T. (1977). Muscle spindle to the human jaw muscles, *Journal of Dental Research*, **56**, 901–909.

Lamarre Y. and Lund, J.P. (1975). Load compensation in human masseter muscles. *Journal of Physiology*, **253**, 21–35.

Lavigne, G., Kim, J. S., Valigette, C. and Lund, J. P. (1987). Evidence that periodontal pressoreceptors provide positive feedback to jaw closing muscles during mastication. *Journal of Neurophysiology*, **58**, 342–358.

Loewenstein, W. R. and Rathkamp, R. (1955). A study on the pressoreceptive sensibility of the tooth. *Journal of Dental Research*, **34**, 287–294.

Lund, J. P., Richmond, F. J. R., Touloumis, C., Patry, Y. and Lamarre, Y. (1978). The distribution of Golgi tendon organs and muscle spindles in masseter and temporalis muscles of the cat. *Neuroscience*, **3**, 259–270.

Lund, J. P. and Matthews, B. (1981). Responses of temporomandibular joint afferents recorded in the Gasserian ganglion of the rabbit to passive movements of the mandible. In Y. Kawamura and Dubner (Eds.), *Oral-Facial Sensory and Motor Functions*. (pp. 153–160), Quintessence: Tokyo.

Maeda, T., Kannari, K., Sato, O. and Iwanaga, T. (1990). Nerve terminals in human periodontal ligament as demonstrated by immunohistochemistry for neurofilament protein (NFP) and S-100 protein. *Archives of Histology and Cytology*, **53**, 259–265.

Miller, M. G. (1981). Trigeminal differentiation and ingestive behavior in rats, *Journal of Comparative and Physiological Psychology*, **95**, 252–269.

Morimoto, T. (1983). Mandibular position sense in man, In Y. Kawamura (Ed.), *Oral sensory mechanisms, Frontiers of Oral Physiology*, Vol. 4. (pp. 80–101). Karger: Basel.

Morimoto, T., Inoue, T., Masuda, Y. and Nagashima, T. (1989). Sensory components facilitating jaw-closing muscle activities in the rabbit. *Experimental Brain Research*, **76**, 424–440.

Morimoto, T., Takebe, H., Sakan, I. and Kawamura, Y. (1978). Reflex action of extrinsic tongue muscles by jaw-closing muscle proprioceptors. *Japanese Journal of Physiology*, **29**, 461–471.

Sakada, S. (1983). Physiology of mechanical senses of the oral structure. In Y. Kawamura (Ed.), Vol. 4 *Frontiers of Oral Physiology*, (pp. 1–32). Karger: Basel.

Schaerer, P., Legault, J. V. and Zander, H. A. (1966). Mastication under anesthesia. *Helvetica Odontologica Acta*, **10**, 130–134.

Sherrington, C. S. (1900). Cutaneous sensations. In *Schafer's Textbook of Physiology* (pp. 920–1001). Pentland: London.

Schoen, R. (1931). Untersuchungen uber Zungen und Kieferreflexe, I mitteilung: der Kiefer-Zungen reflexe und andere proprioceptive Reflex der Zunge und der Kiefermuskulatur. *Archiv experimental Pathologie und Pharmacologie*, **160**, 29–48.

van Steenberghe, D. and de Vries, J. H. (1978). Psychophysical threshold level of periodontal mechanoreceptors in man. *Archives of Oral Biology*, **23**, 41–49.

Taylor, A. (1990). Proprioceptic control of jaw movement. In A. Taylor (Ed.), *Neurophysiology of the Jaws and Teeth* (pp. 237–267). Macmillan: London.

Thilander, B. (1961). Innervation of the temporomandibular joint capsule in man. *Transactions of the Royal School of Dentistry*, **2**, 1–67.

Zeigler, H. P. (1975). Trigeminal differentiation and hunger in pigeon (*Columbia livia*). *Journal of Comparative and Physiological Psychology*, **89**, 827–844.

7

Gustatory control of ingestion

THOMAS R. SCOTT AND BARBARA K. GIZA

Department of Psychology, University of Delaware, Newark, DE 19716, USA

7.1. Introduction

The sense of taste motivates and directs ingestion. Animals use the chemical senses to select foods that support diverse biochemical processes from an often inhospitable environment. Even in human societies where food is abundant, fleeting gustatory pleasure provides an irresistible attraction.

7.2. Bases of taste preference

As taste guides feeding, so does the process of feeding influence taste. The satisfaction derived from taste depends not only on the quality and intensity of the stimulus, but on past experience with that taste and on the chronic and acute physiological needs of the animal.

7.2.1. Experiences

The gustatory experiences of suckling rats form taste preferences that persist into adulthood (Capretta and Rawls, 1974). Preferences also develop through association of a taste or other sensory quality with positive reinforcement (Revusky, 1974), in particular with a visceral reinforcement such as occurs with the administration of a nutrient of which the animal has been deprived (Booth *et al.*, 1974). By pairing an initially aversive flavour with nutritional repletion, an overwhelming preference for that flavour may be established within days (Sclafani, 1991). Indeed, gustatory preferences can even be created in humans and other animals by mere familiarity through constant exposure (Capretta *et al.*, 1973).

However, the most potent effect of experience on subsequent food selection results from the establishment of a conditioned taste aversion (CTA). This is an especially efficient form of conditioning in which an intense aversion may be developed through a single pairing of a novel taste (CS) with gastrointestinal malaise (US) (Garcia *et al.*, 1955). The conditioned aversion to a taste solution is so readily established, so potent and so resistant to extinction that the CTA protocol has itself become a standard tool for studying physiological processes and taste-related behaviour (Bernstein, 1978).

7.2.2. Physiological need

An animal's physiological condition is closely related to its choice of foods. The "body wisdom" demonstrated in rats in cafeteria studies by Richter and in newly weaned infants (Davis, 1928) appeared to be the result of innate taste-directed changes in food-selection. While evidence exists for an innate hunger for sodium (Beauchamp, 1983), for most nutrient deficiencies the adaptive food selections have since been shown to be learned (Rozin, 1976; Gibson and Booth, 1986). Compensatory feeding behaviour has been shown in cases of experimentally induced deficiencies of thiamine (Seward and Greathouse, 1973), threonine (Halstead and Gallagher, 1962) and histidine (Sanahuja and Harper, 1962). Booth (1974) found that elimination of protein from as few as one or two meals was sufficient for amino acid balance to acquire reinforcing properties. It is presumed that the physiological benefits of dietary repletion are paired with the taste or other food cue that preceded those benefits, creating a conditioned preference for that tastant. Since physiological needs are in constant flux, the hedonic value of a taste based on experience of repletion must be quite labile.

A more abiding preference exists for sodium. Evolving in sodium-deficient environments, many mammals seek out and consume salt wherever it is found and, when plentiful, in excess of need. Both rats and human beings select sodium salts in their diets, even when sodium replete (Denton, 1976). This preference becomes exaggerated under conditions of salt deficiency. Humans depleted by pathological states (Wilkins and Richter, 1940) or by experimental manipulations (McCance, 1936; Beauchamp *et al.*, 1990) show an increased preference for salt. Rodents subjected to adrenalectomy (Epstein and Stellar, 1955), dietary restrictions (Cullen and Harriman, 1973) or injections of formalin (Stricker and Wilson, 1970), cyclophosphamide (Mitchell *et al.*, 1974), aldosterone (Wolf and Handel, 1966) or DOCA (Wolf and Quartermain, 1966) show sharp increases in sodium consumption. This compensatory response to the physiological need for salt results from a change in the preference for tasted sodium. Concentrations of salt that had been evaluated negatively and rejected under conditions of sodium repletion evoke a positive hedonic response and acceptance when deprived.

While changes in amino acid or sodium levels may be appreciated over a period of days, the availability of sugars, especially glucose, is of more immediate concern. In just hours of deprivation, the definition of which chemicals are acceptable may be expanded as the dangers of malnutrition weigh against those of toxicity. After only minutes of feeding, the definition may reverse again. Thus the decision to swallow or reject, as well as the hedonic reactions which guide that decision, change with momentary conditions of need. Our common experience joins with the results of psychophysical studies to confirm this logic: with deprivation, foods become more palatable; with satiety, less so (Cabanac, 1971).

7.2.3. Neural bases

This chapter reviews neurophysiological support for three postulates: (1) The basic function of the gustatory code is to differentiate toxins from nutrients. (2) The interaction between taste and ingestion in the rat is largely organized at the hindbrain level. (3) Taste input is modifiable, able to shift from a pattern that elicits ingestive reflexes to one that evokes rejection, or *vice versa*, according to past experience or physiological need. The conundrum of body wisdom may then be focused on the specific issue of how gustatory afferent signals are modified, a process about which little is known.

7.3. Gustatory differentiation of toxins from nutrients

The relationship between the sense of taste and ingestive processes is established at its most fundamental level by the finding that the neural code for taste quality is based on a dimension of physiological welfare.

No issue is more basic to the characterization of a sensory system than a definition of the dimension(s) upon which its perceptions are based. The study of auditory perception, for example, may proceed in a logical fashion because it is known that pitch derives largely from the frequency of the incident pressure waves. It is clear from a study of gustatory receptor mechanisms that taste perceptions are also influenced by the physical characteristics of relevant stimuli. However, these are largely independent for each of the basic tastes. For example, while hydrogen ion concentration may partly account for sourness, it relates not at all to saltiness. This has led to the suggestion that taste is a system of independent modalities, sensitive to unrelated chemical features of the environment.

The introduction of multidimensional scaling techniques into taste research permitted an approach to identifying these modalities independently of assumptions about the system's organization. A wide range of sapid stimuli could be applied to the tongue and the profile of neural or behavioural response elicited by each chemical determined across individual subjects or neurones. Similarities among the profiles could then be measured by correlation coefficients or other statistics and used in a multidimensional scaling programme to generate a spatial representation of relative stimulus similarity. Multidimensional scaling is an iterative data-reduction technique in which coordinates are computed for a set of points such that the distances between pairs of these points are inversely proportional to the measured similarities in the data set. The goodness of this fit is assessed by variance measures such as R-squared (RSQ), stress or the coefficient of alienation.

In the final multidimensional configuration, the higher the correlation between two profiles of activity, the closer the stimuli that evoked those profiles are situated in the space. The axes of the space in which the taste compounds are located represent those stimulus characteristics that underlie gustatory discriminability among that set of compounds. This is because the computer program assigns a stimulus to a position in that space on the assumption that the amounts of each of the characteristics that a stimulus possesses are what determines the taste quality evidenced in the response profile. Therefore, the dimension along which taste quality perception is organized may be determined by finding the optimal match between a stimulus characteristic and each axis of the multidimensional space. The importance of any characteristic in determining taste quality within the set of compounds tested is proportional to the total data variance accounted for by the axis with which it is matched (see also Chapter 1 of this volume).

This approach has now been used to interpret both psychophysical (Schiffman and Erickson, 1971) and electrophysiological (Scott and Mark, 1987) data. The common result of these studies—the latter of which is summarized below— is that a major dimension along which taste stimuli may be organized relates not to any one physicochemical feature of the stimulus molecule, but rather to a physiological characteristic: its effect on the welfare of the organism.

The response profiles that 16 compounds evoked from 42 taste cells in the nucleus tractus solitarius of the rat are placed according to their relative similarities in the multidimensional space that appears in Fig. 7.1. The RSQ (0.95) indicates an excellent fit to the two-dimensional solution in which a preponderance of the variation in distances among compounds is distributed along dimension 1. Thus the stimulus characteristic that corresponds to this dimension is the major factor in permitting these compounds to be distinguished by the taste system. Stimulus placement on this dimension correlates 0.83 ($p < 0.001$) with stimulus toxicity, as indexed by the rat's oral LD_{50} for each compound. Stimulus toxicity, then, provides an excellent basis for predicting relative taste quality across a wide range of chemicals. This dimension is shown in isolation at the bottom of Fig. 7.1. Dimension 2, accounting for much less variance, corresponds to the total response magnitude evoked by each stimulus.

Although the relationship between LD_{50} and stimulus position on the dominant first dimension is highly significant, it does not explain the full discriminative capacity of the rat. The organic acids all generate patterns that correlate about $+0.90$ with that of strychnine, which is a thousand times more toxic and also easily discriminated from the acids. Thus the rat has access to more information than is assayed by this scaling procedure. That additional input may derive from the temporal characteristics of the evoked response. In the analysis above, only the total numbers of spikes that accumulated from each stimulus application during 5 seconds of evoked activity were considered. Yet

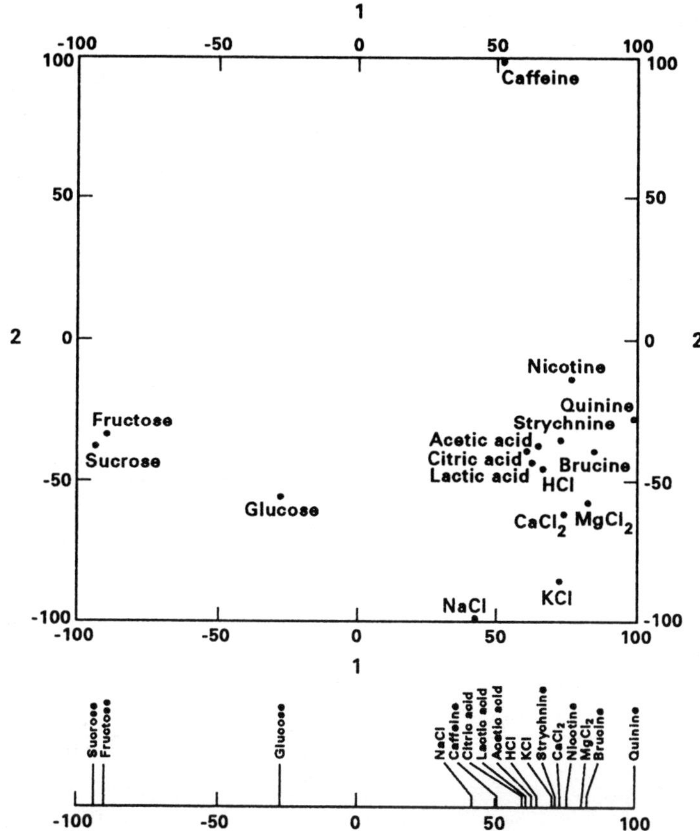

FIG. 7.1. *Top*: A two-dimensional space representing relative similarities among taste qualities as determined by activity profiles across neurones. Position on dimension 1 correlates 0.83 ($p < 0.001$) with stimulus toxicity (rat oral LD_{50}). Position on dimension 2 correlates 0.76 ($p < 0.001$) with total activity across all neurones by each stimulus. *Bottom*: Dimension 1 shown in isolation. (From Scott and Mark (1987) with permission.)

the time course of this accumulation not only carries reliable information regarding taste quality (Nagai and Ueda, 1981) but also may be sufficient to activate appropriate reflexive responses to chemicals in behaving rats. Therefore temporal profiles were generated, each composed of fifty 100-msec bins collapsed across neurones for every stimulus, and the correlation matrix and multidimensional representation were calculated as before. The result is shown at the top of Fig. 7.2. Dimension 1 is again dominant and placement on it correlates $+0.85$ ($p < 0.001$) with oral LD_{50}; it is represented in isolation at the bottom of Fig. 7.2.

This temporal analysis separates strychnine from the organic acids, all of whose temporal discharge patterns correlate only slightly over $+0.40$ with that

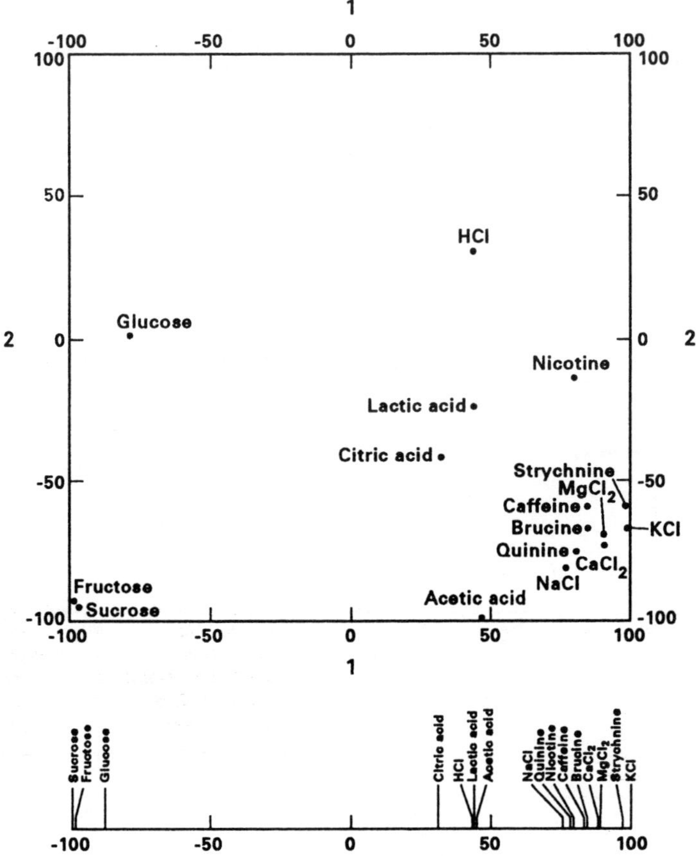

FIG. 7.2.   *Top*: A two-dimensional space representing relative similarities among taste qualities as determined by activity profiles across time. Position on dimension 1 correlates 0.85 ($p < 0.001$) with stimulus toxicity. Position on dimension 2 is undefined. *Bottom*: Dimension 1 shown in isolation. (From Scott and Mark (1987) with permission).

of the alkaloid. It also introduces coefficients that would be perplexing (the temporal pattern of salty NaCl correlates nearly $+ 0.90$ with those of bitter $CaCl_2$ and $MgCl_2$) if the earlier analysis (in which these coefficients reach only around $+ 0.55$) were not available.

Certain stimulus pairs, then, are clearly discriminable from the distribution of response patterns either across neurones or across time. Other pairs (quinine and sucrose) are readily discriminable by either means. When neither factor provides separation, however, as with $MgCl_2$ and $CaCl_2$, the rat cannot easily make a behavioural discrimination. The conclusion is that taste quality information is carried in a spatiotemporal code, both the spatial and temporal aspects of which are organized predominantly along dimensions that relate to the rat's welfare, measured here by stimulus toxicity.

The dimension of physiological welfare underlying the organization of taste quality is not simply the sweet-bitter dichotomy. We believe that it is a more fundamental physiological dimension upon which the psychological perceptions of sweetness and bitterness may be based. Chemicals in the environment have promoted or disrupted physiological functions in animals, providing nutrition or causing illness or death. Selection among foragers has then favoured the taste system that activates the appropriate behaviour or hedonic tone (attraction to nutrients, revulsion by toxins) to match the physiological consequences of ingestion.

The effectiveness of the gustatory neural code in protecting the biochemical welfare of the organism has been demonstrated in a behavioural study in which naive rats were offered free access to the same set of compounds used in the electrophysiological study (Scott, Chang, Bechtel and Mark, 1981, and unpublished). Rats rejected compounds in direct proportion to their toxicity ($r = 0.78$). The more strictly a compound was rejected by rats, the more likely it was to be described by humans as bitter or nauseous; chemicals avidly accepted by rats were labelled sweet or pleasant. Thus the neural dimension of toxicity-nutrition is perceptually coded in the hedonic evaluation of pleasant (mainly sweet) or unpleasant (mainly bitter) sensations. That which tastes good is good for you.

7.4. Hindbrain interaction of taste with ingestion

In rats much of the interaction between taste and ingestion is organized at the level of the hindbrain, as evidenced by anatomical, physiological and behavioural studies.

7.4.1. Anatomical interactions

Hindbrain taste nuclei are in close proximity to regions that receive visceral afferents. They also have reciprocal connections with diencephalic and telencephalic areas that are associated with ingestion, reinforcement and emotion.

(a) Ascending fibres

In the rat, gustatory and visceral afferents both terminate in the nucleus tractus solitarius (NTS) (Rogers and Novin, 1983). Taste input projects to the rostral one-third of the nucleus while axons of the vagus nerve conduct visceral information more caudally. Communication between visceral sensory and gustatory regions of the NTS may occur directly via intrasolitary (Becker and Travers, 1990) projections and indirectly by way of the subjacent reticular formation (Hermann and Rogers, 1982; see also Chapters 2 and 3 of this volume). Moreover, both gustatory and viscerosensory regions of NTS are contiguous with parasympathetic neurones of the dorsal motor nucleus of the vagus, suggesting a path by which taste might mediate such functions as the cephalic phase of insulin release (Powley, 1991). From both regions of NTS, axons project rostrally to the parabrachial nuclei (PBN) where convergence of gustatory and visceral afferents onto single neurones has been shown neurophysiologically (Hermann and Novin, 1980). Bifurcating projections from PBN to the thalamocortical axis and to ventral forebrain structures provide the possibility of parallel processing of taste quality and of hedonic-motivational information, respectively (see Chapter 9).

(b) Descending fibres

The communication between hindbrain taste relays and their forebrain targets is reciprocal. Both the lateral hypothalamus and central nucleus of the amygdala send centrifugal projections to PBN and NTS, completing a loop through which gustatory or visceral signals could be modulated by central states (Saper *et al.*, 1976; Bereiter *et al.*, 1979).

7.4.2. Physiological and behavioural considerations

Evidence has been given above that the hindbrain code for taste is based largely on stimulus toxicity, which relates directly to taste preferences and to feeding behaviour.

In addition, acceptance-rejection reflexes associated with taste stimuli are stereotypical in motor pattern and unaltered by loss of tissue rostral to the brainstem. Steiner and his colleagues have studied orofacial responses to taste stimuli in normal human adults (Hertz *et al.*, 1971; Halbreich and Steiner, 1977), full-term and premature human neonates (Steiner, 1973), anencephalic and hydroanencephalic human neonates, blind adolescents (Steiner, 1976; Steiner and Abraham, 1978), patients with craniofacial abnormalities (Steiner, 1974), retarded humans (Gonshorowitz, 1977) and subhuman primates (Steiner and Glaser, 1985), mammals (Steiner, 1973), and birds (Gentle, 1971). To the extent permitted by their various limitations, subjects across this phylogenetic, ontogenetic and pathological range all reacted similarly to the application of basic taste stimuli. Facial expressions are adaptive both in dealing with the chemical (swallowing if appetitive, clearing the mouth if aversive) and in communicating a hedonic dimension to other members of the species. Moreover, this hedonic monitor is located in the brainstem (Pfaffmann *et al.*, 1977; Steiner and Meiraz, 1977) and is neurally intact by the seventh gestational month in humans (Pfaffmann, 1978; Steiner, 1979).

The orofacial expressions of intact and decerebrate rats during application of diverse taste stimuli support a similar conclusion (Grill and Norgren, 1978a; Grill and Norgren, 1978b). The reactions of decerebrates, though slightly muted, were nearly identical to those of normal rats to all taste qualities. Thus acceptance-rejection behaviours are fully integrated at or caudal to the mid-brain. This conclusion is consistent with the finding of a toxicity dimension in the hindbrain gustatory code, upon which such discriminative behaviour could be based.

As will be discussed in greater detail in the next section, the exogenous administration of factors that influence feeding behaviour alters taste-evoked neural activity in the hindbrain. The responsiveness of taste neurones in NTS is subject to the influence of conditioned taste aversion (Chang and Scott, 1984), sodium deprivation (Jacobs *et al.*, 1988) and several satiety factors including gastric distension (Glenn and Erickson, 1976), elevated blood glucose levels (Giza and Scott, 1983), hyperinsulinaemia (Giza and Scott, 1987a) and glucagon (Giza *et al.*, 1989).

Finally, factors that influence feeding are effective even when their administration is restricted to the hindbrain. When glucose availability in the hindbrain was reduced through use of the metabolic inhibitors 5-thioglucose (Ritter *et al.*, 1981) or phlorizin (Flynn and Grill, 1985) or by deprivation in decerebrate rats (Grill and Norgren, 1978c; Flynn and Grill, 1983), the animals became hyperphagic. Decerebrate rats also demonstrate satiety (Grill, 1980) and anorexia induced by the gut peptide cholecystokinin (Grill and Smith, 1988). Therefore, the caudal brainstem has some capacity to regulate ingestive functions in a homeostatic manner. It is not, however, fully competent in this regard. Decerebrates do not eat spontaneously, nor do they respond behaviourally to the challenges of sodium depletion (Grill *et al.*, 1986), dehydration (Grill and Miselis, 1981) or aversive conditioning (Grill and Norgren, 1978c). Interestingly, the preabsorptive insulin release elicited by the taste of glucose, and eliminated by the development of a conditioned aversion to glucose in intact rats, is also eliminated in decerebrate preparations by pairing the taste of glucose with intestinal malaise. This, combined with the observation that taste responses are altered in the NTS by conditioning (cf. following section), implies that a conditioned taste aversion leaves its trace in the hindbrain even though it does not achieve a behavioural manifestation.

In sum, the hindbrain possesses sensory monitors for toxins and nutrients from the external environment, and can guide orofacial expressions associated with the analyses performed by these monitors. Moreover, chemical signals from the internal milieu may be integrated with those from the mouth. Gustatory and visceral sensory signals are modified by changes in physiological condition (see below), and several homeostatic processes are organized in the hindbrain, fully independent of forebrain involvement.

7.5. Modification of taste input by experience or need

In the preceding sections we argued that the taste system differentiates toxins from nutrients and that this is accomplished low in the neuraxis but induces powerful hedonic experiences. While providing a broad and effective system for maintaining the biochemical welfare of the species, this organization would not recognize the idiosyncratic intolerances or needs of the individual or be sensitive to changes in those needs over time. To serve these requirements, the taste code must be plastic.

7.5.1. Aversion conditioning

The neural substrates of conditioned taste aversions (CTAs) have been investigated in scores of experiments, most of which have involved ablating selected structures and testing the ability of subjects to retain former CTAs or to develop subsequent aversions. Many brain areas have been implicated in aversion learning, but only amygdaloid and hypothalamic participation seems unequivocal. Rarely have recordings been made from neurones of conditioned animals to determine the effects of a CTA on taste-evoked activity. Aleksanyan *et al.* (1976) reported that the preponderance of hypothalamic activity evoked by saccharin in rats shifted from the lateral to the ventromedial nucleus with formation of a saccharin CTA. DiLorenzo (1985) recorded the responses evoked by a series of taste stimuli in the pontine parabrachial nucleus in rats, then paired the taste of NaCl with gastrointestinal malaise and repeated the recordings. The response to NaCl increased significantly and selectively in a subset of gustatory neurones.

Chang and Scott (1984) recorded single-unit gustatory-evoked activity from the NTS of three groups of rats: unconditioned (exposed to the taste of the saccharin CS with no induced nausea), pseudoconditioned (experienced only the US, nausea, with no gustatory cue), and conditioned (taste of saccharin CS paired with nausea). Comparisons were performed among the groups' responses to an array of 12 stimuli, including the saccharin CS, a more concentrated saccharin solution, and sugars, salts, acids and an alkaloid, through which alterations in the entire gustatory code resulting from this taste-learning experience could be evaluated. The CS evoked a significantly larger response from conditioned animals than from those of either control group, and the effect was limited to the 30% of neurones that showed a sweet-sensitive profile of responsiveness. Temporal analyses of the activity evoked from this subgroup of saccharin-sensitive neurones revealed that nearly the entire increase in discharge rate was attributable to a burst of activity that diverged from control group levels 600 msec following stimulus onset, peaked at 900 msec and returned to control levels by 3000 msec (Fig. 7.3). Thus the major consequence of the conditioning procedure was to increase responsiveness to the saccharin CS through a well-defined peak of activity. The same enhanced response and temporal pattern were evoked to a lesser extent by other sweet stimuli (fructose, glucose, sucrose), providing a likely neural counterpart to generalization of the aversion.

Since a range of taste stimuli was used in this study, the effect of an increased response to the CS and related chemicals could be evaluated in terms of taste

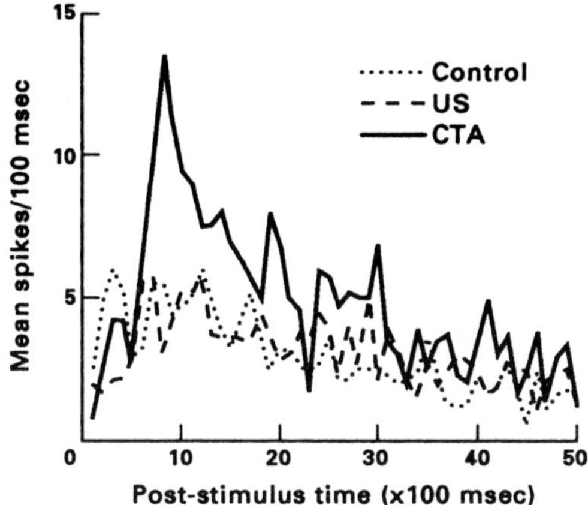

FIG. 7.3.   Post-stimulus time histogram to 0.0025 M sodium saccharin calculated from the 13 most responsive neurones in each group of rats. Groups are the same as in Fig. 7.4. (From Chang and Scott (1984), with permission.)

quality. The pattern of activity evoked by saccharin was altered by the conditioning procedure. How did the new pattern relate to those of other compounds? Correlation matrices and multidimensional spaces were constructed (as described earlier) in three dimensions from the response profiles of unconditioned and conditioned rats (Fig. 7.4). In the former group (Fig. 7.4A), representing a normally functioning taste system, the stimulus arrangement is similar to that seen by others (Doetsch and Erickson, 1970). The basic distinction between sweet and non-sweet compounds is apparent, as is the precise arrangement of stimuli within the non-sweet set: the four sodium or lithium salts should be and are virtually indistinguishable by the patterns they evoke. The complex sweet-salty-bitter taste of concentrated (0.25 M) sodium saccharin is represented appropriately between the sweet and non-sweet clusters. The consequence of the conditioning procedure is to disrupt this clear organization (Fig. 7.4B). The relative similarity among non-sweet qualities is reduced and the sharp distinction between sweet and non-sweet is blurred. Moreover, the relative increase in similarity between sweet and non-sweet chemicals is differential, with the greatest decrease in distance being between the saccharin CS and bitter quinine. The rearrangement is so decisive that quinine is nearly as close to the CS as it is to the acids, a finding that would be quite aberrant in a normally functioning taste system. This may offer a neural concomitant to the increase in similarity between behavioural reactions evoked by quinine and sweet compounds to which an aversion has been conditioned (Grill and Norgren, 1978a).

These findings reinforce the reports by Aleksanyan *et al.* (1976) and DiLorenzo (1985) that the responses of taste neurones in the brainstem and indeed hindbrain are subject to modification by experience. Moreover, a condi-

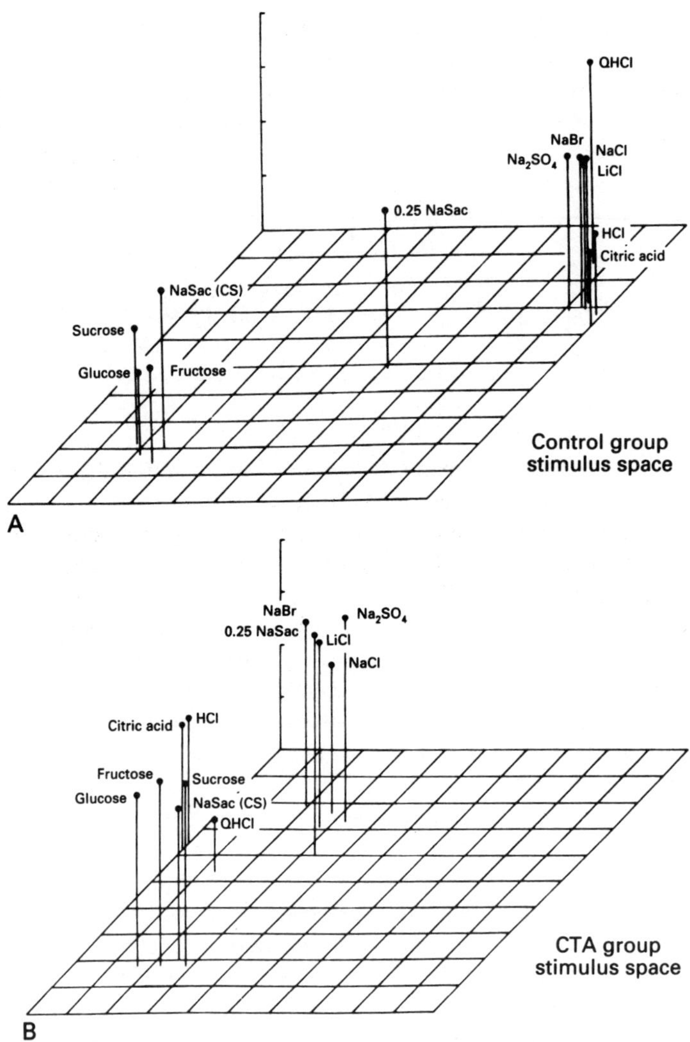

FIG. 7.4. Three-dimensional spaces representing relative similarities among stimuli as determined from activity profiles across neurones in the control group (A) and the CTA group (B). Na Saccharin (CS) is at the same coordinates in each space. (From Chang and Scott (1984), with permission.)

tioned aversion, which in humans affects primarily the preference for a CS rather than its perceived quality, causes a rearrangement of the taste space derived from neurophysiological responses of the NTS in rats. This suggests again that the hindbrain neural code in rats combines sensory discrimination with a hedonic component.

7.5.2. *Physiological need*

High concentrations of NaCl that had been rejected under conditions of sodium repletion are avidly accepted when the rats are sodium-depleted. This change had been thought to follow from a decreased gustatory sensitivity to salt. Contreras (1977) analysed the responses of single fibres in the chorda tympani nerve to NaCl in replete and in sodium-deficient rats. Depletion was accompanied by a specific reduction in salt responsiveness in those 40% of fibres that were most sodium-responsive. This decreased sensitivity was seen as mediating the observed shift in the acceptance curve to higher concentrations.

This intensity-based interpretation has recently been revised. Recording centrally, from taste neurones in the NTS, Jacobs *et al.* (1988) confirmed a moderate overall reduction in responsiveness to sodium in salt-deprived rats. Separate analyses of activity among different neurone types, however, revealed that the responsiveness of the salt-sensitive group of cells was profoundly depressed and that this effect was partly offset by a sharp increase in activity among sweet-sensitive cells. The net effect was to transfer the burden of coding sodium from salt- to sweet-sensitive neurones. This implies a change not so much in perceived intensity as in perceived quality in sodium-deficient rats: salt should now taste "sweet" or perhaps "good" to such an animal. Multidimensional spaces based on the responses of replete and of sodium-deprived rats confirm shifts in the neural codes for sodium and lithium salts toward those of sucrose and fructose (Fig. 7.5). This concurs with the eagerness with which deprived rats consume sodium, an avidity usually reserved for sugars.

The hedonic appeal of foods rises with caloric deprivation and falls with satiety. These effects are also accompanied by alterations in gustatory afferent activity.

The responsiveness of gustatory receptor cells and peripheral taste nerves was reduced in frogs by gastric distension and enhanced by food deprivation (Dua-Sharma *et al.*, 1973; Sharma and Doss, 1973). All modulatory effects were lost with bilateral transection of the glossopharyngeal nerve, implying that they are mediated through efferent fibres in peripheral taste nerves. Accordingly, Hellekant (1971, 1972) suggested that receptor inhibition results from the alterations in efferent activity of the chorda tympani that he found to be induced by gastric distension in rats. Another link in the neural chain that binds physiological state to afferent responsiveness may be the vagus nerve, whose transection reduces or abolishes these phenomena (Sharma *et al.*, 1977).

The effects of deprivation or satiety extend to the CNS. Glenn and Erickson (1976) recorded multiunit activity from the NTS of freely fed rats and noted a pattern of differential modification similar to that seen at the periphery. Gastric distension selectively depressed activity, with the greatest effect on that evoked by sucrose, followed in descending order by that to NaCl, HCl and finally quinine, the responses to which were unmodified. Relief from distension reversed the effect over a 45-minute period. When rats were deprived of food for

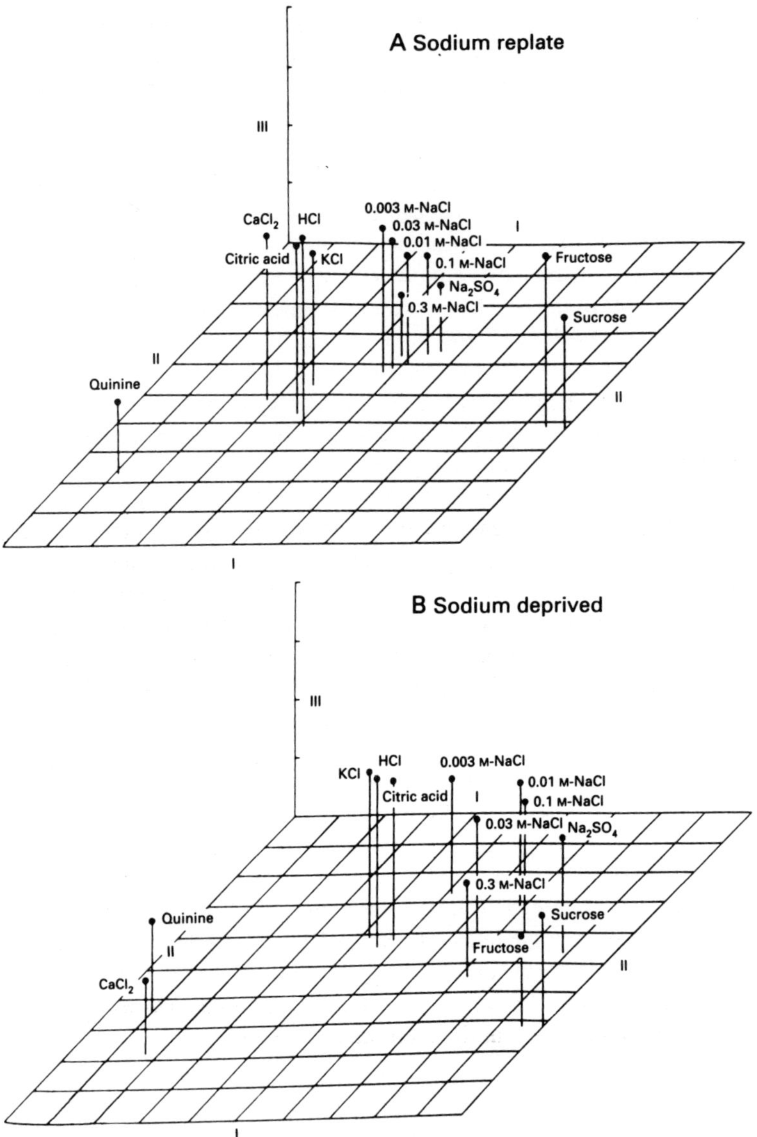

FIG. 7.5.   Three-dimensional spaces representing relative similarities among stimuli as determined from neural responses in sodium-replete (A) and sodium-deprived (B) rats. (Reprinted from Jacobs *et al.* (1988) with permission.)

48–72 hours, however, the influence of distension on taste was lost, suggesting that the modulating processes are subject to the overall nutritive state of the animal.

The effects of other satiety factors on taste-evoked activity in the NTS have now been investigated. Giza and Scott (1983) recorded multiunit responses to taste stimuli before and after intravenous loads of 0.5 g/kg glucose or a vehicle. The glucose infusion caused a significant reduction in gustatory responsiveness to glucose, with a maximum effect occurring 8 min following the intravenous load. Recovery took place over 60 minutes as blood glucose approached normal levels. Responsiveness to NaCl and HCl was suppressed to a lesser degree and for briefer periods, while quinine-evoked responses were unaffected. Similarly, both intravenous administration of insulin at a dose of 0.5 U/kg (Giza and Scott, 1987a) and an intrahepatic portal vein infusion of glucagon at 40 µg/kg (Giza *et al.*, 1989) resulted in a transient suppression of NTS activity evoked by the taste of fructose or glucose. Thus, procedures that deliver utilizable calories to the tissues and glucose to the brain result in depression of food intake and are associated with reductions in the afferent activity evoked by preferred tastes. These findings imply that the pleasure that sustains ingestion is reduced, making termination of the meal more likely.

If taste activity in NTS is influenced by the rat's nutritional state, then intensity judgements may change with satiety. To examine intensity perception in the rat, conditioned taste aversions were formed to 1.0 M glucose and the degree of behavioural generalization to this and a range of other concentrations was measured (Scott and Giza, 1987). Hyperglycaemic rats reacted to all glucose concentrations 50% less strongly than did conditioned animals with no glucose load (Giza and Scott, 1987b). Since the learned aversion should be relatively specific to 1.0 M glucose, the neural suppression in the hindbrain that results from hyperglycaemia may be manifested in the perception of reduced intensity of sweet compounds at all strengths.

The corresponding neurophysiological studies in primates have given different results (cf. the chapter by E. T. Rolls in this volume). Taste-evoked activity in the macaque's NTS is not suppressed by a glucose load nor indeed do people report a decline in perceived intensity with satiety. However, satiety does inhibit activity in the second cortical taste relay in the orbitofrontal cortex of the macaque, implying that the hedonic appreciation thought to be mediated in this region is related to physiological need. Therefore related processes are at work in the rat and monkey, but at different neural levels.

7.6. Conclusions

The taste system lies at the interface between discrimination and digestion. It combines the qualities of rapid stimulus identification and spatiotemporal coding that are associated with the exteroceptive senses with subsequent processes of recognition in the still-mysterious codes of the visceral senses. Its basic organization, inherited from ancestors who valued the tastes of nutrients over those of toxins, arrays the chemical world along a dimension of physiological welfare. The afferent signals from nutrients drive feeding reflexes and activity in forebrain areas that mediate reward, while those of toxins activate withdrawal reflexes and forebrain neurones associated with aversion. Thus a first approximation of body wisdom is inherited through taste.

The plasticity of the system permits finer adjustments. The appropriateness of accepting a food must be verified by its physiological consequences. If a taste that arouses feeding reflexes and reward is followed even after some hours by nausea, the gustatory signal in the rat's hindbrain is altered—by as yet unknown mechanisms—toward one of a toxin. Onward from the neural level of that alteration (the hindbrain in a rat or the forebrain in primates), nothing need change, for the afferent code now activates reflexes and emotions associated with a poison. Conversely, an inherently aversive or neutral taste that is paired with nutritional repletion gains hedonic appeal. The rat pup with garlic in its mother's milk, the Italian child with tomato sauce on his pasta, the Japanese with ginger on her rice, each develops a preference for the flavours that precede prandial satisfaction. Thus bodily wisdom is extended from the level of species to that of the culture and of the individual.

Finally, the gustatory afferent signal, with its consequences for pleasure or revulsion, is modifiable over the course of days, hours or even minutes by transient needs. With nutritional deficiency, responsiveness to taste is heightened, as is the appeal of foods. Repletion brings a suppression of the gustatory signal for reward from that taste, and yet other compounds may still arouse pleasure. It follows that a diet selected purely according to the reinforcing value of its components will be a varied one, perhaps skewed this way or that by the age or metabolic idiosyncracies of the individual. That variability need not reside in hindbrain reflexes or forebrain reward systems, but can be manifested in the form of the gustatory code that drives them.

The wisdom of the body is in what arouses pleasure at the moment and that in turn is a consequence of alterations in the neural code for tastes. In this way, the taste system reduces the diverse and frequently hostile environment to a chemical subset that effectively satisfies the complex and changing requirements of the internal milieu.

References

Aleksanyan, A. A., Buresova, O. and Bures, J. (1976). Modification of unit responses to gustatory stimuli by conditioned taste aversion in rats. *Physiology and Behavior*, **17**, 173.

Beauchamp, G. K., Bertino, M., Burke, D. and Engelman, K. (1990) Experimental sodium depletion and salt taste in normal human volunteers. *American Journal of Clinical Nutrition*, **51**, 881–889.

Beauchamp, G. K., Bertino, M. and Engelman, K. (1983). Modification of salt taste. *Annals of Internal Medicine*, **98**, 763–769.

Becker, D. and Travers, S. (1990). Projections of electrophysiologically-identified loci in the orally-responsive solitary nucleus. *Society for Neuroscience Abstracts*, **16**, 404.

Bereiter, D. A., Berthoud, H. R. and Jeanrenaud, B. (1979). Oropharyngeal and hypothalamic input to the same NTS neurons. *Society for Neuroscience Abstracts*, **5**, 125.

Bernstein, I. L. (1978). Learned taste aversions in children receiving chemotherapy. *Science*, **200**, 1302–1303.

Booth, D. A. (1974). Acquired sensory preferences for protein in diabetic and normal rats. *Physiological Psychology*, **28**, 344–348.

Booth, D. A., Stoloff, R. and Nicholls, J. (1974). Dietary flavor acceptance in infant rats established by association with effects of nutrient composition. *Physiological Psychology*, **2**, 313–319.

Cabanac, M. (1971). Physiological role of pleasure. *Science*, **173**, 1103–1107.

Capretta, P. J., Moore, M. J. and Rossiter, T. R. (1973). Establishment and modification of food and taste preferences: Effects of experience. *Journal of General Psychology*, **89**, 27–46.

Capretta, P. J. and Rawls, L. H. (1974). Establishment of a flavor preference in rats: Importance of nursing and weaning experience.*Journal of Comparative and Physiological Psychology*, **86**, 670–673.

Chang, F-C. T. and Scott, T. R. (1984). Conditioned taste aversions modify neural responses in the rat nucleus tractus solitarius. *Journal of Neuroscience*, **4**, 1850–1862.

Contreras, R. (1977). Changes in gustatory nerve discharges with sodium deficiency: A single unit analysis. *Brain Research*, **121**, 373–378.

Cullen, J. W. and Harriman, A. E. (1973). Selection of NaCl solutions by sodium-deprived Mongolian gerbils in Richter-Type drinking tests. *Journal of Psychology*, **83**, 315–321.

Davis, C. M. (1928). Self selection of diet by newly weaned infants: An experimental study. *American Journal of Diseases of Children*, **36**, 651–679.

Denton, D. A. (1976). Hypertension: A malady of civilization? In M. P. Sambhi, (Ed.), *Systemic Effects of Antihypertensive Agents* (pp. 577–583). Stratton Intercontinental Medical Books: New York.

DiLorenzo, P. M. (1985). Responses to NaCl of parabrachial units that were conditioned with intravenous LiCl. *Chemical Senses*, **10**, 438.

Doetsch, G. S. and Erickson, R. P. (1970). Synaptic processing of taste-quality information in the nucleus tractus solitarius of the rat. *Journal of Neurophysiology*, **33**, 490–507.

Dua-Sharma, S., Sharma, K. N. and Jacobs, H. L. (1973). The effect of chronic hunger on gustatory responses in the frog. *Physiologist*, **16**, 300.

Epstein, A. N. and Stellar, E. (1955). The control of salt preference in adrenalectomized rat. *Journal of Comparative and Physiological Psychology*, **46**, 167.

Flynn, F. W. and Grill, H. J. (1983). Insulin elicits ingestion in decerebrate rats. *Science*, **221**, 188–189.

Flynn, F. W. and Grill, H. J. (1985). Fourth ventricular phlorazin dissociates feeding from hyperglycemia in rats. *Brain Research*, **341**, 331–336.

Garcia, J., Kimmeldorf, D. J. and Koelling, R. A. (1955). Conditional aversion to saccharin resulting from exposure to gamma radiation. *Science*, **122**, 157–158.

Gentle, M. J. (1971). Taste and its importance to the domestic chicken. *British Poultry Science*, **12**, 77–86.

Gibson, E. L. and Booth, D. A. (1986). Acquired protein appetite in rats: dependence on a protein-specific need state. *Experientia*, **42**, 1003–1004.

Giza, B. K., Deems, R. O., VanderWeele, D. A. and Scott, T. R. (1989). Glucagon administration affects taste sensitivity. *Appetite*, **12**, 212.

Giza, B. K. and Scott, T. R. (1983). Blood glucose selectively affects taste-evoked activity in rat nucleus tractus solitarius. *Physiology and Behavior*, **31**, 643–650.

Giza, B. K. and Scott, T. R. (1987a). Intravenous insulin infusions in rats decrease gustatory-evoked responses to sugars. *American Journal of Physiology*, **252**, R994–1002.

Giza, B. K. and Scott, T. R. (1987b). Blood glucose level affects perceived sweetness intensity in rats. *Physiology and Behavior*, **41**, 459–464.

Glenn, J. F. and Erickson, R. P. (1976). Gastric modulation of gustatory afferent activity. *Physiology and Behavior*, **16**, 561–568.

Gonshorowitz, J. (1977). Facial expressions in response to taste and food-related odor stimuli in the mentally retarded. Unpublished M.D. Thesis, Hebrew University.

Grill, H. J. (1980). Production and regulation of ingestive consummatory behavior in the chronic decerebrate rat. *Brain Research Bulletin*, **5**, 79–87.

Grill, H. J. and Miselis, R. R. (1981). Lack of ingestive compensation to osmotic stimuli in chronic decerebrate rats. *American Journal of Physiology*, **240**, R81.

Grill, H. J. and Norgren, R. (1978a). The taste reactivity test: I. Mimetic responses to gustatory stimuli in neurologically normal rats. *Brain Research*, **143**, 263–279.

Grill, H. J. and Norgren, R. (1978b). The taste reactivity test. II. Mimetic responses to gustatory stimuli in chronic thalamic and chronic decerebrate rats. *Brain Research*, **143**, 281–297.

Grill, H. J. and Norgren, R. (1978c). Chronically decerebrate rats demonstrate satiation but not bait shyness. *Science*, **201**, 267–269.

Grill, H. J., Schulkin, J. and Flynn, F. W. (1986). Sodium homeostasis in chronic decerebrate rats. *Behavioral Neuroscience*, **100**, 536–543.

Grill, H. J. and Smith, G. P. (1988). Cholecystokinin decreases sucrose intake in chronic decerebrate rats. *American Journal of Physiology*, **254**, R853–R856.

Halbreich, U. and Steiner, J. E. (1977). An interaction between hyperbaric pressure and taste in man. *Archives of Oral Biology*, **22**, 287–289.

Halstead, W. C. and Gallagher, B. B. (1962). Autoregulation of amino acids intake in the albino rats. *Journal of Comparative and Physiological Psychology*, **55**, 107–111.

Hellekant, G. (1971). The effect of stomach distension on the efferent activity in the chorda tympani nerve of the rat. *Acta Physiologica Scandinavia*, **83**, 527–531.

Hellekant, G. (1972). Influences on the impulse pattern in efferent chorda tympani nerve fibers in the rat. In D. Schneider (Ed.), *Olfaction and Taste IV*, (pp. 308–315). Wissenschaftliche Verlagsgesellschaft MBH: Stuttgart.

Hermann, G. and Novin, D. (1980). Morphine inhibition of parabrachial taste units reversed by naloxone. *Brain Research Bulletin*, **5**, (Suppl. 4), 169–173.

Hermann, G. E. and Rogers, R. C. (1982). Hepatic and gustatory interactions in the brainstem of the rat. *Society for Neuroscience Abstracts*, **8**, 200.

Hertz, D., Steiner, J. E., Zuckerman, H. and Pizanyi, S. (1971). Psychological and physical symptom formation in menopause. *Psychotherapy and Psychosomatic Medicine*, **19**, 47–52.

Jacobs, K. M., Mark, G. P. and Scott, T. R. (1988). Taste responses in the nucleus tractus solitarius of sodium-deprived rats. *Journal of Physiology*, **406**, 393–410.

McCance, R. A. (1936). Experimental sodium chloride deficiency in man. *Proceedings of the Royal Society of London Series B*, **119**, 245–268.

Mitchell, D., Parker, L. F. and Woods, S. C. (1974). Cyclophosphamide-induced sodium appetite and hyponatremia in the rat. *Pharmacology, Biochemistry, and Behavior*, **2**, 627–630.

Nagai, T. and Ueda, K. (1981). Stochastic properties of gustatory impulse discharges in rat chorda tympani fibers. *Journal of Neurophysiology*, **45**, 574–592.

Pfaffmann, C. (1978). The vertebrate phylogeny, neural code, and integrative processes of taste. In E. C. Carterette and M. P. Friedman (Eds.), *Handbook of Perception*, Vol. VIA. (pp. 51–123). Academic Press: New York.

Pfaffmann, C., Norgren, R. and Grill, H. J. (1977). Sensory affect and motivation. *Annals of the New York Academy of Science*, **290**, 18–34.

Powley, T. L. and Berthoud, H.-R. (1991). Neuroanatomical bases of cephalic phase reflexes. In M. I. Friedman (Ed.), *Chemical Senses: Appetite and Nutrition*, (pp. 391–404). Marcel Dekker, Inc.: New York.

Revusky, S. H. (1974). Retention of a learned increase in the preference for a flavored solution. *Behavioral Biology*, **11**, 121–125.

Ritter, R. C., Slusser, P. G. and Stone, S. (1981). Glucoreceptors controlling feeding and blood glucose: Location in the hindbrain. *Science*, **213**, 451–452.

Rogers, R. C. and Novin, D. (1983). The neurological aspects of hepatic osmoregulation. In A. N. Epstein (Ed.), *The Kidney in Liver Disease*, (pp. 337–350). Elsevier Science Publishing Co.: New York.

Rozin, P. (1976). The selection of foods by rats, humans and other animals. In Rosenblatt, R. A. Hinde, C. Beer and E. Shaw (Eds.), *Advances in the Study of Behavior*. Vol. 6, (pp. 21–76). Academic Press: New York.

Sanahuja, J. C. and Harper, A. E. (1962). Effect of amino acid imbalance on food intake and preference. *American Journal of Physiology*, **202**, 165–170.

Saper, C. B., Loewy, A. D., Swanson, L. W. and Cowan, W. M. (1976). Direct hypothalamo-autonomic connections. *Brain Research*, **177,** 305–312.

Schiffman, S. S. and Erickson, R. P. (1971). A psychophysical model for gustatory quality. *Physiology and Behavior*, **7,** 617–633.

Sclafani, A. (1991). Nutrient-based conditioned flavor preferences. In E. D. Capaldi and T. Powley (Eds.), *Taste, Experience and Feeding*, (pp. 139–156).

Scott, T. R., Chang, F-C. T., Bechtel, S. L. and Mark, G. P. (1981). Unpublished observations.

Scott, T. R. and Giza, B. K. (1987). A measure of taste intensity discrimination in the rat through conditioned taste aversions. *Physiology and Behavior*, **41,** 315–320.

Scott, T. R. and Mark, G. P. (1987). The taste system encodes stimulus toxicity. *Brain Research*, **414,** 197–203.

Seward, J. P. and Greathouse, S. R. (1973). Appetitive and aversive conditioning in thiamine-deficient rats. *Journal of Comparative Physiology*, **83,** 157–167.

Sharma, K. N. and Doss, M. J. K. (1973). Excitation and control of gustatory chemoreceptors. *Proceedings of the 10th International Conference of Medical and Biological Engineering*, **3,** 53.

Sharma, K. N., Jacobs, H. L., Gopal, V. and Dua-Sharma, S. (1977). Nutritional state/taste interactions in food intake: Behavioral and physiological evidence for gastric/taste modulation. In M. R. Kare and O. Maller (Eds.), *The Chemical Senses and Nutrition*. Academic Press: New York.

Steiner, J. E. (1973). The gustofacial response: Observation on normal and anencephalic newborn infants. In J. F. Bosma (Ed.), *Symposium on Oral Sensation and Perception IV*, (pp. 254–278). NIH-DHEW: Bethesda.

Steiner, J. E. (1974). Testing of the senses of taste and smell in craniofacial abnormalities. Paper presented at the annual meeting of the American Cleft-Palate Association, Boston.

Steiner, J. E. (1976). Further observations on sensory motor coordinations induced by gustatory and olfactory stimuli. *Israel Journal of Medical Science*, **12,** 1231.

Steiner, J. E. (1979). Human facial expressions in response to taste and smell stimulation. In H. W. Reese and L. Lipsett (Eds.), *Advances in Child Development*, (pp. 257–295). Academic Press: New York.

Steiner, J. E. and Abraham, F. E. (1978). Gustatory and olfactory functions in patients affected by Usher's syndrome. *Chemical Senses*, **3,** 93–98.

Steiner, J. E. and Glaser, D. (1985). Orofacial motor behavior-patterns induced by gustatory stimuli in apes. *Chemical Senses*, **10,** 452.

Steiner, J. E. and Meiraz, O. (1977). Food-odor-preference and human masticatory performance. *Israel Journal of Dental Medicine*, **26,** 31–38.

Stricker, E. M. and Wilson, N. E. (1970). Salt-seeking behavior in rats following acute sodium deficiency. *Journal of Comparative and Physiological Psychology*, **72,** 416–420.

Wilkins, L. and Richter, C. P. (1940). A great craving for salt by a child with corticoadrenal insufficiency. *Journal of the American Medical Association*, **114,** 866–868.

Wolf, G. and Handel, P. J. (1966). Aldosterone-induced sodium appetite: Dose response and specificity. *Endocrinology*, **78,** 1120–1124.

Wolf, G. and Quartermain, D. (1966). Sodium chloride intake of desoxycorticosterone-treated and of sodium-deficient rats as a function of saline concentration. *Journal of Comparative and Physiological Psychology*, **61,** 288–291.

8

Olfactory processing controlling food and fluid intake

RÉMI GERVAIS

Laboratoire de Physiologie Neurosensorielle, URA CNRS 180,
Université Claude Bernard Lyon I, 69622 Villeurbanne Cedex, France

ALTHOUGH the information provided by food and fluid odours is not predictive of the metabolic benefit of intake, it is decisive in whether the animal will be attracted or repelled. Olfactory cues give information on the location and the quality of food which is critical to the initiation phase of intake. After physical contact is established between the animal's mouth and the food, gustatory and other oral cues provide further information on palatability which ultimately modulates the speed of consumption and the size of the meal.

This schematic description of the role of olfaction in the control of intake leads to two important points. Taking a decision whether or not intake should be initiated requires both fine olfactory discrimination and also comparison of the incoming information with prior olfactory experiences. The neural bases of these two processes are therefore a matter of, first, the coding of olfactory information and, secondly, the mechanisms involved in olfactory learning and memory. Our present knowledge of these neural processes will be described in the present chapter, but after we have considered behavioural data that illustrate the importance of olfaction in the organization of food and fluid intake.

8.1. Behavioural importance of olfaction in food and fluid intake

8.1.1. Inborn responses

Several examples of inborn preferences with a low level of individual variability, mediated by the chemical senses, have been observed in animals and humans (for reviews, see Le Magnen, 1971; Schaal, 1988). The recording of facial expressions in babies just after birth in responses to odours as well to tastes has been claimed to demonstrate innate reactions (Steiner, 1979). In rabbit pups,

119

it is well established that nipple-search behaviour requires olfactory information emanating from the lactating doe. Interestingly, the ability of maternal pheromone to elicit nipple search behaviour and ultimately milk intake does not require post-natal experience. The typical behaviour is observed in the first day of life and pups' performances in milk intake during the short (3 min) single daily meal did not differ on day 5 between pups raised normally and those deprived of any contact with their mother (Hudson and Distel, 1986).

Clear-cut spontaneous reaction to odours have also been observed in adults. In 1953, Barnett and Spencer showed that rats exhibit strong ingestive preferences among unfamiliar foods: whole wheat was preferred to barley grains and to white wheat. Since such preferences involve tactile, gustatory and olfactory cues, further experiments were conducted to evaluate the specific participation of the sense of smell. For example, the rat was given a choice between two samples of synthetic diet, one flavoured by citral and the other flavoured by eucalyptol. The rats clearly preferred to eat the citral form, thus revealing a spontaneous olfactory preference (Le Magnen, 1959).

However, showing inborn reactions to odours does not solve the difficult problem of the origins of such response, that is to say whether it results from genetic constraints or from *in utero* experience. There is no reason to exclude genetically programmed responses, but unambiguous demonstration of such determinants is difficult to obtain. Moreover, as pointed out by Schall (1988), pre- and post-natal experience are in functional continuity. This is illustrated by several lines of experimental evidence that young animals display behavioural responses to odours determined by *in utero* experience. Injection of odorous substances (e.g. apple juice, orange juice) into the amniotic fluids of pregnant rats modify behavioural responses postnatally (Smotherman, 1982; Stickford *et al.*, 1982). In a typical experiment, apple juice extract was injected at day 20 of gestation, followed by injection of lithium chloride as an aversively conditioning stimulus. Ten days after birth, the presence of the odour of apple juice in the environment reduced the attraction of conditioned animals towards an anaesthetized lactating female (Smotherman, 1982). Such experiments strongly suggest that both peripheral and central olfactory pathways are functional *in utero*, an hypothesis which is in accordance with numerous anatomical data (for review, see Brunjes and Frazier, 1986), a metabolic mapping study (Pederson *et al.*, 1983) and the behavioural response of the foetus to odours (Smotherman and Robinson, 1988).

8.1.2. *Learned responses*

Human alimentary habits are largely governed by cultural and familial influences, thus illustrating the importance of learning in smell and taste preferences. In experimental animals, the role of experience in food selection has been extensively studied, most often by using neophobia and conditioned aversion (Bures and Buresova, 1977).

One common way to study neophobia is first to accustom the animals to eating a single daily meal on a standard diet. On the test day, the usual food is perfumed with an unfamiliar odour and the amplitude of the reduction in the amount eaten is a measured of the strength of response to that novel stimulus. Another procedure is based on a two-choice test in which the usual food is presented simultaneously with a potentially palatable new food. In this case, the animal invariably prefers the usual food.

Repeated daily presentation of the new food is accompanied by a reduction of the neophobic response and its disappearance in 3 to 4 days (Royet and Pager, 1980). Olfactory information has been shown to be primarily involved in both of these kinds of test situation. Animals rendered anosmic by destruction of the mucosal odour receptors or the olfactory bulbs ate significantly more novel food than control animals did (Bures and Buresova, 1977; Royet, 1983). Thus, it is likely that the smell of an unknown source of food or fluid in natural conditions induces more or less pronounced avoidance behaviour.

To induce a conditioned aversion, the presentation of a new food is paired with a noxious stimulus. Within minutes following presentation of the novel food, the animal receives an intraperitoneal injection of substance inducing nausea (e.g. LiCl or apomorphine). A single pairing produces a marked and long-lasting repulsion from the novel food. This conditioned response involves primarily gustatory cues, but olfactory information again is of importance (Royet, 1983).

8.1.3. Neural mechanisms

So, to whatever extent responses to odours are inborn or learned postnatally, olfactory cues play a critical role in food and fluid selection throughout the animal's life. We will now examine neural mechanisms which could explain, at least partly, how olfactory information is used in order to generate the behavioural response.

8.2. Anatomical organization of olfactory pathways

In mammals, olfactory information reaches the brain via two independent routes. They are the main and the accessory olfactory systems (reviews by Holley and MacLeod, 1977, and Price, 1987).

For the main system (Fig. 8.1), receptor cells are found in posterior turbinates of nasal cavity and their axons reach the main olfactory bulb (MOB). Output neurones of MOB (mitral and tufted cells) send most of their projections to the anterior olfactory nucleus (AON), the anterior and posterior parts of pyriform cortex and to the periamygdaloid cortex. Recent data suggest that some axons also reach the supraoptic nucleus (Smithson *et al.*, 1989).

Receptor cells of the accessory system are found in a cigar-shaped structure, the vomeronasal organ, situated in the rostral portion of nasal cavity. These

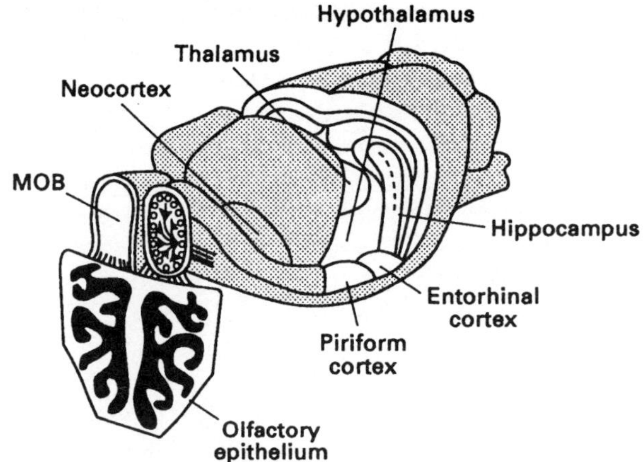

FIG. 8.1.   Schematic representation of the rat brain and the main structures involve in the processing of the olfactory information, LOT, lateral olfactory tract; MOB, main olfactory bulb. (Drawing from A. Holley.)

receptors project exclusively to the accessory olfactory bulb (AOB), a miniature bulb embedded in the dorsal, posterior part of the MOB. The main target structure of AOB relay neurones is the cortico-medial nucleus of the amygdala.

The dual organization of main and accessory systems still persists at higher levels. Efferents from pyriform cortex reach the lateral entorhinal area, mediodorsal thalamus and orbito-frontal neocortical areas. Since the entorhinal area massively projects to the hippocampus through the perforant path, the hippocampal formation is just two synapses away from the olfactory receptors. Thus, information provided by the main system is mainly addressed to the limbic area and to neocortical structures known to be critically involved in learning, memory and cognitive functions. In contrast, the periamygdaloid cortical relay of the accessory system sends fibres which run into the stria terminalis to the ventromedial hypothalamus and the medial preoptic area. Thus, the ultimate site of action of the accessory system is the hypothalamo-hypophyseal axis (Keverne, 1983; Wysocki and Meredith, 1987).

One important characteristic of olfactory connections is the existence of centrifugal fibres (for reviews, see Holley and MacLeod, 1977; Scott and Harrison, 1987). As observed in other sensory systems, olfactory structures are most often reciprocally connected. For instance, each structure receiving axons from the output neurones of the MOB projects back to the bulb. Thus, the MOB receives innervation from AON, the whole extent of pyriform cortex, the nucleus of the lateral olfactory tract and the anterior hypothalamus. In addition, several other areas send projections to the bulb but do not receive monosynaptic projections from it. They are the well-known ascending modulatory systems using noradrenaline, serotonin and acetylcholine as transmitters (Halasz and Shepherd,

1983). Noradrenergic and serotonergic cell bodies are found exclusively in the brain stem, in locus coeruleus and raphe nuclei respectively. Cholinergic neurones belong to the anterior part of the basal cholinergic complex of the telencephalon. Consequently, MOB activity is likely to be determined by peripheral as well as central influences.

Existence of such well-differentiated main and accessory systems raises the question of their respective participation in behavioural responses to the odours of foods and fluids. At least two sets of data strongly support a major involvement of the main system. First, behavioural responses to food odour result from perception of distant chemical information. It is well established that volatile airborne molecules mediating such information mainly activate receptor cells of the main olfactory system. Access of odorant molecules to the vomeronasal organ requires physical contact between the animal's nose and the source. Secondly, as predictable from the organization of central projections, the main system deals with the perception and recognition of odours from past experience, while activation of accessory system leads to neuroendocrine reflex responses (Lloyd-Thomas and Keverne, 1982).

8.3. Principles of olfactory coding

Understanding the neural mechanisms supporting control of food and fluid intake by odours requires basic knowledge on information coding in the main olfactory system. Most data have been collected at the first two levels of the system, the neuroreceptors and the olfactory bulb (reviews Holley and MacLeod, 1977; Scott and Harrison, 1987; Holley, 1991). Neuroreceptor cells have been electrically recorded mostly in anaesthetized or curarized frogs, while olfactory bulb activity has been recorded in frogs and anaesthetized or awake mammals.

In brief, it is now well established that each neuroreceptor cells is fairly broadly tuned in the sense that it is likely to respond to several different pure chemical compounds. Since the subset of neuroreceptor cells sensitive to a given compound seems non-homogeneously distributed in the epithelial sheet, it has been proposed that each chemical generates a spatial pattern of activity in the neuroreceptor population (chemotopy). Another important feature is that neural representations of different chemical compounds are likely to share in common numerous receptor cells; in other words, epithelial maps for different odours overlap. However, what remains unknown is how the brain manages the instability of each map which is likely to be induced by breathing. In fact, even if the respiratory rate remains constant, the distribution of airborne molecules in the nasal cavity is unlikely to be the same from one inhalation to the next. This variability should further increase when respiration rate changes from the resting level to sniffing. Thus, central processing of olfactory information must ensure stability of perception in spite of large variability of the input. Based on experimental data collected in numerous experiments, computer simulation and

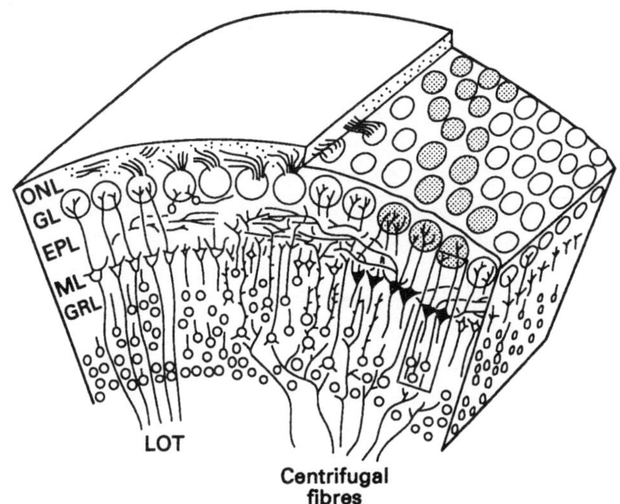

FIG. 8.2.   Schematic representation of the organization of the mammalian main olfactory bulb (MOB). Hatched glomeruli and black mitral cell bodies illustrate how an odour is thought to activate cellular elements of the structure. Abbreviations for layers from the top to the bottom: ONL, olfactory nerve layer; EPL, external plexiform layer; GL, glomerular layer; ML, mitral cell body layer; GL, granule cell layer; LOT, lateral olfactory tract. (Drawing from A. Holley.)

theoretical considerations, Skarda and Freeman (1987) proposed that stability is ensured as early as the level of the MOB.

In the main olfactory bulbs about 1000 neuroreceptor cell axons converge on each relay neurone (mitral and tufted cells). They synapse in a well-defined anatomical structure, the olfactory glomerulus (100 µm diameter). This projection is topographically organized, although far from being point to point (Astic and Saucier, 1986; Saucier and Astic, 1986). By using ^{14}C 2-deoxyglucose (2-DG) as a metabolic marker, it was shown that each odour compound induced activity in a reproducible subset of olfactory glomeruli (Jourdan, 1981; Royet *et al.*, 1987). Since, in MOB of mammals, the unique primary dendrite of each second-order neurone arborizes in a single glomerulus, it is reasonable to assume that the chemotopy reflected at the glomerular level is transferred to the layer of relay neurones, an assumption supported by recent electrophysiological experiments (Buonviso and Chaput, 1990) (Fig. 8.2). This suggests that one or a few adjacent glomeruli, together with underlying relay neurones to which they are connected, constitute a columnar functional organization. However, in spite of the fact that the principle of columnar functioning has been proposed in most sensory cortical areas, what is observed at the macroscopic level is still unclear in function (Swindale, 1990).

Examination of behavioural effects of lesions illustrates this problem. If the 2-DG activity generated at the glomerular level by presentation of an odour represents one aspect of the neural code, one could expect that partial or total destruction of the map would impair detection and discrimination. Surprisingly, partial but large lesions of the MOB were unable to reveal such perturbations (Hudson and Distel, 1987; Slotnick *et al.*, 1987). For example, destruction of more than 80% of the bulbar map of a complex odour (the perfume Chanel No. 5) used in a conditioning experiment in rabbit pups had no detectable effect on the behavioural response of animals to this odour (Hudson and Distel, 1987). This suggests that higher structures are very efficient in recognizing "distorted" bulbar maps.

Most of the data we have examined in this section were not obtained in the context of food intake behaviour. Nevertheless, they are of importance in understanding how food odours can be used by animals in the natural environment, mainly because they underline the fact that variability in the neural representation of the stimulus at the levels of the receptors and the bulb has no major influence on odour recognition. This is of significance, because plants and animals are sources of very complex odours that are likely to vary over time, as for example from one season to another. Over a broad range of variation, animals should be able to maintain accurate recognition of food odour sources and so produce appropriate behavioural responses.

8.4. Recognition of complex odours

Most experimental investigations of olfactory coding use pure chemical compounds such as isoamyl acetate, camphor, eucalyptol, etc. Since food odours are chemically complex stimuli, the question arises how such information is analysed. One basic point to clarify is whether perception of biological odours is analytic or involves *Gestalt* properties. In a set of experiments directly addressing this question (Staubli *et al.*, 1987), thirsty rats were trained to discriminate pairs of odours to obtain water as a reward. Each session consisted of 20 trials where one odour of the pair was consistently associated with reward, while the other was not. Each odorant was a mixture of 3 to 4 pure chemicals, while each pair of odours presented had several components in common (e.g. ABC vs ABD or ABCD vs ABCE), a condition which is likely to occur with food odours—for example, two different meats or two different plants. As might be predicted from natural behaviour, the rats solved each task fairly rapidly, independently of the odours' complexity and similarity in composition. Moreover, when discrimination criterion had been achieved, the rats had to solve a task where the odours were the elements unshared by the pair of complex stimulus (i.e. C vs D or D vs E). It was clear that the rats did not recognize these one-component odours. This provides strong support for *Gestalt* properties of olfactory discrimination.

However, in a different experimental procedure where odours were substituted by multisite electrical stimulation of the olfactory bulb, animals succeeded in discriminating the stimulation of one site which had previously been stimulated together with several others (Mouly *et al.*, 1985). This suggests that the olfactory system has some analytical capacity at least when components of a mixture are tested individually. However, this situation does not occur frequently in natural conditions. Hence, the perception of food and fluid odours is likely to be global in character. This is in accordance with the view that olfaction is not used to provide detailed analysis of the different components of food, which theoretically might be used to predict metabolic benefit. Rather, olfaction gives information on location and palatability.

8.5. Functional duality in olfactory pathways

Output neurones from the MOB reach central structures via two different routes. On one hand, axons of relay neurones join in the most lateral aspect of the bulb to form the lateral olfactory tract (LOT). LOT fibres innervate anterior and posterior parts of pyriform cortex and the lateral entorhinal area. Hence, lateral bulb output sends information to the pyriform-thalamoneocortical pathway and to the entorhinal-hippocampal system. On the other hand, mitral and tufted cell axons and their collaterals also project to the anterior olfactory nucleus (AON) and the anterior hypothalamic area. Moreover, the AON is itself a source of projections to the medial forebrain bundle (MFB) and lateral hypothalamic area (Broadwell, 1975; Barone *et al.*, 1981). Thus, medial bulb output is a source of information to diencephalic structures involved in the specific activation of ingestion (Le Magnen, 1987) and through MFB to mesencephalic and brainstem areas critical for general arousal of behaviour (Morgane, 1969). A series of behavioural experiments was designed to test the hypothesis that the lateral pathway could sustain discriminative olfactory perception while the medial pathway could mediate arousing properties of odours (Pager, 1977).

Behavioural responses to food odours were observed in hungry animals in three different situations: (a) rats had to find the source of a food odour (Gervais and Pager, 1982); (b) rats were presented an unknown food (neophobic response); (c) rats were presented with a familiar food 24 hours after aversion had been conditioned to it (Royet and Pager, 1982). Another study used natural odorants that elicited spontaneous emotional responses (Cattarelli, 1982)— for instance the odour of a natural predator, the fox, which induces a freezing reaction in rats.

These four behavioural responses were studied in intact rats and following selective lesions to lateral or medial olfactory pathways. It was observed that animals in which only medial pathways remained intact (anterior limb of anterior commissure and MFB) were poor at finding the food odour and in displaying neophobic and emotional responses to odours. However, when the odour had acquired a strong arousing effect following aversion conditioning, the olfactory aversion remained. Intact lateral projections alone were able to sustain nearly normal behavioural responses.

So, data obtained in awake animals showed that the lateral pathway can sustain recognition of odours, while the medial cannot. However, data obtained from presentation of odours during slow wave sleep led to the opposite conclusion. Animals with the medial pathway intact only were wakened by odours as often as normal rats. For instance, the awakening power of fox odour remained higher than that of other biological odours, while food odour retained its capacity to induce cortical desynchronization more often in hungry animals than in satiated animals. The difference in effect of food odour between hungry and satiated rats indicated efficient integration of olfactory information to internal cues. In contrast, destruction of the medial pathway abolished the awakening effect of odours (Cattarelli, 1982; Gervais and Pager, 1982). So, odour recognition during slow wave sleep selectively involves the medial pathway.

Overall, these experimental data confirmed the general hypothesis that medial and lateral olfactory pathways are functionally complementary. Normal utilization of olfactory information requires the lateral pathway in wakefulness and the medial pathway in slow wave sleep. This is strong evidence for quite different central processing of information.

8.6. Central responses to food and fluid odours

As a first step toward understanding the neural basis of behavioural reponses to food odours, several authors have examined how external and internal cues are integrated at different levels of the olfactory pathways (Le Magnen, 1987). Most data have been collected from the first relay structure of the system, the olfactory bulb. Those obtained in behaving animals clearly demonstrate that the response of the olfactory bulb to food odours largely depends on the learned value of the odour and the nutritional state of the animal at the time of presentation (Pager, 1986a). This is possible because the olfactory bulb receives projections from numerous central structures. Thus, integration of internal and external cues is achieved as early in processing as the olfactory bulb. Knowing that, it is not surprising that similar electrophysiological correlates are found at more central levels. Most data collected from neocortical, thalamic, limbic and hypothalamic areas do indeed give evidence of integration between olfactory information and internal cues.

8.6.1. Neocortical, thalamic and limbic areas

The most coherent set of data regarding olfactory processing in thalamo-neocortical pathways has been obtained in Old World monkeys (Takagi, 1984). The stimuli used were pure compounds including those having fruity odours, such as undecalactone (peach-like odour) and isoamyl acetate (banana odour). Unit recordings revealed the existence of two adjacent neocortical olfactory areas in the orbito-frontal (OF) area, the lateral and posterior (LPOF) area and the centro-posterior portion (CPOF). The LPOF showed one of the highest level of selectivity to odours observed anywhere in the brain, with more than 50% of the cells responding to only one of the 8 odours used. Neurones in the CPOF showed an entirely different response pattern, similar to that observed in the mediodorsal thalamus (MD). In both MD and CPOF, the majority of neurones responded to 4–6 odours. These observations led the authors to propose a role of LPOF in odour discrimination, while CPOF could integrate odour sensations.

In the limbic area, most recording has been done in the hippocampal formation. Rats had to perform a discrimination between two cue-odours while unit recordings were made from presumptive relay neurone in the CA1 region of the dorsal hippocampus (Eichenbaum *et al.*, 1987; Weiner *et al.*, 1989). Some cells fired selectively according to which odours were presented (pure olfactory coding), the configuration of odour cues (place and olfactory coding), the locus of the behavioural response (place and task coding) or a combination of these variables. So, hippocampal cells integrate information provided by most sensory modalities including olfaction and the spatial coordinates of the animals. Thus, hippocampal processing is likely to be important for cognitive aspects of smell related to experience and location.

The amygdala is another major component of the limbic system, but its reactivity to odours is still poorly understood. In the anaesthetized rat (Cain and Bindra, 1972) and in the awake Old World monkey (Takagi, 1984), a large proportion of neurones responded to several odours, including food odours. On this basis, the authors hypothesized that the amygdala is unlikely to play a major role in fine olfactory discrimination. What remains to be done is the identification of behavioural situations in which responses occur.

8.6.2. The hypothalamic area

The importance of the lateral hypothalamic area (LHA) in the control of food intake has been known for many years (Morgane and Jacobs, 1969). Because hypothalamic neurones are sensitive to signals of liver metabolism and gut contents (see Chapters 1–5), as well as input from odours, they are likely to be involved in the differential processing of odour cues according to the metabolic status of the animal.

The first demonstration of an odour-induced response in the LHA was obtained in anaesthetized rats when stimulated by sexual odours (Pfaff and Pfaffmann, 1969,1970; Pfaff and Gregory, 1971; Komisaruk and Bayer, 1972). Recent investigations have presented food odours to behaving animals. In one study, single unit activity in the LHA of the rat was recorded while the animal learned to discriminate chemosensory cues (Nakamura *et al.*, 1989), including solutions having taste only, smell only or taste and smell, prepared from orange and grape extracts. Responses to the smell alone had shorter latencies than responses to the taste alone, but the average values (220 msec and 710 msec respectively) reflect the involvement of polysynaptic circuitry in both cases. Interestingly, neural activity in response to solutions having smell plus taste was the sum of the responses to taste alone and smell alone. This indicates

"Sherringtonian" integrative properties of LHA neurones for chemosensory information. It also suggests that these neurones do not distinguish between taste and odour.

An illustration of the integration of chemosensory information and internal cues is provided by unit recordings in behaving monkeys where responsiveness was examined to odours and to local administration of glucose (Karadi *et al.*, 1989). Different olfactory stimuli evoked responses from 88% of the glucose-sensitive (GS) neurones and 52% of the glucose-insensitive (GIS) neurones. Firing of GIS and GS neurones occurred preferentially at different moments in the operant task (bar pressing, the intake of the reward ...). The GS neurones could integrate multiple chemosensory inputs with both of two endogeneous sources of control over feeding, whereas the GIS cells distinguished among fewer, more specific cues controlling food-acquisition behaviour.

Another hypothalamic region could play an important role in chemosensory analysis governing fluid intake. Sites in the lateral preoptic area of the rat, where intraventricular injection of angiotensin II enhances multiunit activity, are also responsive to odours (Malmo and Malmo, 1988). The authors suggest that convergence of olfactory cues and the angiotensin II signal on the lateral (but not medial) preoptic area might be important in the initiation of fluid intake.

8.6.3. The olfactory bulb

Although there is a lack of data on pyriform cortex, there are numerous observations on responsiveness to food odours at the level of olfactory bulb. This structure conveniently protrudes in front of rodent brain and presents a well-laminated cellular organization. In particular, one major subtype of output neurones, the mitral cells, have their cell bodies delineating a nearly monocellular layer, generating activity at much higher amplitude than neighbouring layers. Thus, neural recording of mitral cells is greatly facilitated.

Moulton (1968) was the first to record mitral cell activity in behaving rats. It was done from indwelling electrodes by telemetry. Available water was signalled to rats by isoamyl acetate. MOB responses to isoamyl acetate were enhanced when the rat responded correctly. This observation suggested that responsiveness could be determined by the behavioural significance of odours.

This view was later reinforced by an investigation showing that multiunit mitral cell responses were enhanced when food odours were presented to hungry rats when compared those obtained in food-satiated rats (Pager *et al.*, 1972). In those experiments, the rats were fed in a single 2-hour daily meal; they were considered to be the hungry state just before the daily meal and in the satiated state just after the meal. The most significant difference between hungry and satiated situations was not the relative amplitude of the neural response but its probability of occurrence on repetitive presentation (Pager, 1986a). For example, 100 presentations of the food odour to hungry rats elicited neural activation in 85 cases while this value fell to around 50 in the same animals when satiated.

This difference in bulbar reactivity to food odour between hungry and satiated situations was called "nutritional modulation".

Analysis of mitral cell unit responses in similar conditions revealed other features of olfactory processing (Pager 1986b). The rats received a puff of food odour followed by delivery of one food pellet. In some cases, the animal ate the food pellet promptly; in other cases the reward was neglected. The striking observation was that the mitral cell response to the odour was biased in a positive direction as the readiness to eat increased. In other words, mitral cell responses to the odour might be used to predict the behavioural outcome (Pager, 1986b).

Other observations support this view. Unit mitral cell activity was recorded in freely behaving rabbits who had to slide a door in order to obtain an odorous food reward (Karpov, 1980). Many of units fired selectively following closure of the door even when the reward was not present. The author therefore proposed that recorded activity reflected not simply olfactory coding but also consequences of the behavioural response.

So it is likely that fairly complex mechanisms of integration operate in the olfactory bulb. For a few hundred milliseconds following arrival of the odour at the receptor cell, what is observed through bulbar electrical activity is a combination of neural correlates of coding processes and centrifugal controls. Hence this structure, although it could be considered to be fairly "peripheral", presents together neurophysiological correlates of sensory coding, motivation and motor responses in the rat (cf. Scott and Giza, Chapter 7, on taste responses in rats).

It was important to demonstrate that the nutritional modulation is under the influence of somatic factors known to be involved in the control of food intake. This was done by manipulating experimentally a single internal parameter characterizing the hungry state or the satiated state. For instance, stomach distension was produced in food-deprived animals to mimic food intake (Chaput and Holley, 1967) or insulin was injected intraperitoneally to produce hypoglycaemia in food-satiated rats (Pager *et al.*, 1972). It was observed that stomach distension in hungry rats reproduced the neurophysiological effect of satiation in the MOB. Conversely, insulin injection in satiated rats massively enhanced bulbar responsiveness to food odour. In both cases, mitral cell responsiveness to nonalimentary odours was unaffected. Thus, internal factors have a direct influence on bulbar reactivity to food odour, a phenomenon which must involve centrifugal fibres to the MOB.

If this bulbar modulation is behaviourally relevant, mitral cell responsiveness should be modified in parallel to the change in the nutritional meaning of odours. So what will happen to the bulbar response to a nonalimentary stimulus that has been repeatedly associated with food intake and to a food odour following the conditioning of aversion?

Eucalyptol elicits a similar, fairly low level of bulbar activation in hungry and satiated rats, even when it has been added to the diet of young rats for 30 days. However, in rats that had received the flavoured diet (already given to the mothers) from weaning to the test period, eucalyptol behaved as a food odour at the mitral cell level (Pager, 1974). This finding further extends the observations on prenatal olfactory learning described at the start of this chapter.

Conversely, the effect of conditioned aversion was studied in adult animals habituated to consuming the eucalyptol-flavoured diet (EF). Before the conditioning aversion, the rate of mitral cell responses to eucalyptol was higher in hungry rats than in satiated rats. A single pairing of intake of EF with treatment producing nausea modified the bulbar pattern of responsiveness; the odour of EF then elicited the same high rate of positive response in satiated as well as in hungry situations (Pager and Royet, 1976).

These two experiments thus illustrated a close parallelism between the behavioural meaning of food odours and mitral cell responses.

Another step in the study of the nutritional modulation of olfactory bulb activity was the attempt to identify the centrifugal pathways involved. Lesions either in the region of the anterior limb of the anterior commissure or in the lateral olfactory tract gave results that indicated that fibres reach the bulb via the medial portion of olfactory peduncle (Pager, 1978). Physiological and anatomical considerations led to the suspicion that these fibres belonged to the medial forebrain bundle. This bundle is heterogeneous, including noradrenergic fibres originating from the locus coeruleus and serotonergic fibres from the raphé nuclei. The participation of these two neurochemically distinct pathways was tested by selective depletion of noradrenaline or serotonin restricted to MOB, by local injection of neurotoxic agents. Recordings of bulbar responses revealed that the nutritional modulation requires participation of noradrenergic but not of serotonergic fibres (Gervais *et al.*, 1988). So, it is likely that, in the olfactory bulb, as in other telencephalic cortical structures, noradrenergic ascending fibres mediate the effect of general arousal and attentiveness which results from perception of relevant information.

Further experiments on the noradrenergic system showed that its action not only results in increased probability of mitral responses to food odours in hungry animals but also allows long-lasting changes to occur following olfactory learning. The activation of bulbar beta-adrenergic receptors at the time of olfactory learning is required for the expression of the long-lasting morphological, metabolic and electrophysiological correlates of olfactory learning that are observed in the bulb of rat pups (Sullivan *et al.*, 1989) and of adult rabbits (Gray *et al.*, 1986).

The fact that neural plasticity is observed in MOB following olfactory learning is intriguing. Plasticity might represent consequences of learning or, alternatively, a prerequisite for learning to be achieved. Recent behavioural experiments support the second possibility: treatment expected to conteract the expression of bulbar plasticity resulted in impairment of both olfactory learning (Lévy *et al.*, 1990) and retention of information over a 5-day period (Mouly *et al.*, 1993). These data were not obtained in the context of food intake control, but it is likely that MOB circuits play an important role in learned response to food odours as well.

8.7. Concluding perspectives

The importance of smell in food and fluid selection is now fully demonstrated, although further investigations are required to refine our knowledge on the respective roles of olfactory and gustatory senses and the determinants of any inborn responses. Understanding the underlying neural mechanisms, however, remains a major challenge.

This field of research offers promising working models because it allows the study of neural correlates in a behavioural context that is highly significant for the animal's survival. Sensory physiologists are aware that our current comprehension of olfactory coding mainly derives from experiments using pure compounds. Studying the neural representation of food odours could provide a more realistic view of the functioning of the olfactory system. It would be fruithful to compare neural maps of food odours obtained in naive animals to those obtained after repetitive ingestion of the source of the odour. Indeed, Freeman and co-workers have recorded electrical signals collected through an array of electrodes positioned at the surface of MOB in awake rabbits, and observed that resting activity and response to a neutral pure compound were significantly modified following pairing odorant presentation with positive or negative reinforcement (Skarda and Freeman, 1987). This again suggests the importance of learning in the functioning of olfactory structures, in the sense that the neural representation of an odour depends more on the animal's experience than on the physico-chemical characteristics of the stimulus.

Finally, it is worth pointing out that the extensively investigated phenomenon of conditioned aversion presents features which are suitable for investigating some poorly understood aspects of the neural basis of learning and memory. One potential productive feature is the relative long delay (tens of minutes) that can occur between the presentation of the cue to be conditioned and the toxic effect of the unconditioned stimulus (e.g. LiCl or apomorphine) to which it is to be associated (Royet, 1983; Holder and Garcia, 1986). One has to postulate some type of "buffer" holding in memory the olfactory and gustatory characteristics of the ingested food. Identification of structures and processes supporting "buffer" or "transient" olfactory memory would represent a major insight into mechanisms of information storage and mental imagery. A second feature is that conditioned aversion is extremely resistant to decay over time, lasting for months and years. In experimental research, this type of learning thus offers a suitable model for studies on the neurobiology of long-term retention of information over a time scale similar to human memory.

References

Astic, L. and Saucier, D. (1986). Anatomical mapping of the neuroepithelium projections to the olfactory bulb in the rat. *Brain Research*, **16**, 445–454.

Barnett, S. A. and Spencer, M. M. (1953). Experiments on the food preferences of wild rat. *Journal Hygenia (London)*, **51**, 16–34.

Barone, F. C., Wayner, M. J., Scharoun, S. L., Guevara-Aguilar, R. and Aguilar-Baturoni,

H. V. (1981). Afferents connections to the lateral hypothalamus: a horseradish peroxidase study in the rat. *Brain Research Bulletin*, **7**, 75–88.

Broadwell, R. R. (1975). Olfactory relationship of the telencephalon and diencephalon in the rabbit. II. An autoradiographic and horseradish peroxidase study of the efferent connections of the anterior olfactory nucleus. *Journal of Comparative Neurology*, **164**, 389–410.

Brunjes, P. C. and Frazier, L. L. (1986). Maturation and plasticity in the olfactory system of vertebrates. *Brain Research Reviews*, **11**, 1–46.

Buonviso, N. and Chaput, M. A. (1990). Response similarity to odours in olfactory bulb output cell presumed to be connected to the same glomerulus: electrophysiological study using simultaneous single-unit recordings. *Journal of Neurophysiology*, **63**, 447–454.

Bures, J. and Buresova, O. (1977). Physiological mechanisms of conditioned food aversion. In N. W. Milgram, L. Kramer and T. M. Alloway (Eds.), *Food Aversion Learning* (pp. 219–255). Plenum Press: New York.

Cain, D. P. and Bindra, D. (1972). Response of amygdala single units to odours in the rat. *Experimental Neurology*, **35**, 98–110.

Cattarelli, M. (1982). The role of the medial olfactory pathways in olfaction: behavioural and electrophysiological data. *Behavioral Brain Research*, **6**, 339–364.

Chaput, M. and Holley, A. (1967). Olfactory bulb responsiveness to food odor during stomach distension in the rat. *Chemical Senses*, **2**, 189–201.

Eichenbaum, H., Kuperstein, M., Fagan, A. and Nagode, J. (1987). Cue-sampling and goal-approach correlates of hippocampal unit activity in rats performing an odor discrimination task. *Journal of Neuroscience*, **7**, 716–732.

Gervais, R. and Pager, J. (1982). Functional changes in waking and sleeping rats after lesions in the olfactory pathways. *Physiology and Behavior*, **29**, 7–15.

Gervais, R. Holley, A. and Keverne, B. (1988). The importance of central noradrenergic influences on the olfactory bulb in the processing of learned olfactory cues. *Chemical Senses*, **13**, 3–12.

Gray, C., Freeman, W. J. and Skinner, J. E. (1986). Chemical dependencies of learning in the rabbit olfactory bulb: acquisition of the transient spatial-pattern change depends on norepinephrine. *Behavioral Neuroscience*, **100**, 585–596.

Halasz, N. and Shepherd, G. M. (1983). Neurochemistry of the olfactory bulb. *Neuroscience*, **10**, 579–619.

Holder, M. D. and Garcia, J. (1986). Role of temporal order and odor intensity in taste-potentiated odor aversions. *Behavioral Neuroscience*, **101**, 158–163.

Holley, A. and Mac Leod, P. (1977). Transduction et codage de l'information olfactive chez les vertébrés. *Journal de Physiologie (Paris)*, **73**, 725–848.

Holley, A. (1991). Neural coding of olfactory information. In T. V. Getchell (Ed.), *Smell and Taste in Health and Disease* (pp. 329–343) Raven Press: New York.

Hudson, R. and Distel, H. (1986). Olfactory-guidance of nipple-search behavior in newborn rabbits. In W. Breighpohl (Ed.), *Ontogeny of Olfaction* (pp. 243–254). Springer-Verlag; Berlin.

Hudson, R. and Distel, H. (1987). Regional autonomy in the peripheral processing of odor signals in newborn rabbits. *Brain Research*, **421**, 85–94.

Jourdan, F. (1981). Spatial dimension in olfactory coding: a representation of the 2-deoxyglucose pattern of glomerular labeling in the olfactory bulb. *Brain Research*, **240**, 341–344.

Karadi, Z., Oomura, Y., Nishino, H. and Aou, S. (1989). Olfactory coding in the monkey lateral hypothalamus: behavioral and neurochemical properties of odor responding neurons. *Physiology and Behavior*, **45**, 1249–1257.

Karpov, A. P. (1980). Analysis of neuron activity in the rabbits olfactory bulb during food-acquisition behaviour. In R. F. Thompson, L. H. Hicks and B. Shvyrkov (Eds.), *Neural Mechanisms of Goal-directed Behavior and Learning* (pp. 273–282). Academic Press: New York.

Keverne, E. B. (1983). Pheromonal influences on the endocrine regulation of reproduction. *Trends in Neuroscience*, **6**, 381–384.

Komisaruk, B. R. and Bayer, C. (1972). Response of diencephalic neurons to olfactory bulb stimulation, odor and arousal. *Brain Research*, **36,** 153–170.

Le Magnen, J. (1959). Efficacité des divers stimuli alimentaires dans l'établissement et la commande de l'appétit chez le rat blanc. *Journal de Physiologie (Paris)*, **51,** 987–998.

Le Magnen, J. (1971). Olfaction and nutrition. In L. M. Beidler (Ed.), *Handbook of Sensory Physiology*, Vol IV, part 1 (pp. 466–482). Springer-Verlag: Berlin.

Le Magnen, J. (1987). Central processing of sensory information in the control of feeding. In D. Ottoson (Ed.), *Progress in Sensory Physiology* vol. 8, (pp. 95–128). Springer-Verlag: Berlin.

Lévy, F., Gervais, R., Kindermann, V., Orgeur, P. and Piketty, V. (1990). Importance of beta-adrenergic receptors in the olfactory bulb of sheep for recognition of lambs. *Behavioral Neuroscience*, **104,** 464–469.

Lloyd-Thomas, A. and Keverne, E. B. (1982). Role of the main and accessory olfactory system in the block to pregnancy in mice. *Neuroscience*, **4,** 907–913.

Malmo, R. B. and Malmo, H. P. (1988). Effects of intraventricular angiotensin II and olfactory stimuli on multiple unit activity in preoptic area and anterior hypothalamic areas: medio-lateral comparison. *Electroencephalography and Clinical Neurophysiology*, **70,** 256–269.

Morgane, P. J. and Jacobs, H. L. (1969). Hunger and satiety. *World Review of Nutrition and Diet*, **10,** 100–213.

Moulton, D. G. (1968). Electrophysiological and behavioral responses to odor stimulation and their correlates. *Olfactologia*, **9,** 69–75.

Mouly, A. M., Vigouroux, M. and Holley, A. (1985). On the ability of rats to discriminate between microstimulations of the olfactory bulb in different locations. *Behavioural Brain Research*, **17,** 45–58.

Mouly, A. M., Kindermann, U., Gervais, R. and Holley, A. (1993). Involvement of the olfactory bulb in consolidation processes associated with long-term memory in rats. *Behavioral Neuroscience*, **107,** 451–457.

Nakamura, K., Ono, T., Tamura, R., Indo, M., Takashima, Y. and Kawasaki, M. (1989). Characteristics of rat lateral hypothalamic neuron responses to smell and taste in emotional behavior. *Brain Research*, **491,** 15–32.

Pager, J., Giachetti,I., Holley, A. and Le Magnen, J. (1972). A selective control of olfactory bulb excitability in relation to food deprivation and satiety in rats. *Physiology and Behavior*, **9,** 573–579.

Pager, J. (1974). A selective modulation of the olfactory bulb electrical activity in relation to the learning of palatability in hungry and satiated rats. *Physiology and Behavior*, **12,** 189–195.

Pager, J. and Royet, J. P. (1976). Some effects of conditioned aversion on food intake and olfactory bulb electrical responses in the rat. *Journal of Comparative and Physiological Psychology*, **90,** 67–77.

Pager, J. (1977). The regulation of food intake behavior: some olfactory central correlates. In J. Le Magnen and P. Mac Leod (Eds.), *Olfaction and Taste VI* (pp. 135–142). IRL Press: London.

Pager, J. (1978). Ascending olfactory information and centrifugal influences contributing to a nutritional modulation of the rat mitral cell responses. *Brain Research*, **140,** 251–269.

Pager, J. (1986a). Neural correlates of odor-guided behaviors. *Experientia*, **42,** 250–256.

Pager, J. (1986b). Unit responses changing with behavioural outcome in the olfactory bulb of unrestrained rats. *Brain Research*, **289,** 87–98.

Pederson, P. E., Stewart, W. B., Greer, C. A. and Shepherd (1983). Evidence for olfactory function *in utero. Science*, **221,** 478–479.

Pfaff, D. W. and Pfaffmann, C. (1969). Olfactory and hormonal influence on the basal forebrain of the male rat. *Brain Research*, **15,** 137–156.

Pfaff, D. W. and Pfaffmann, C. (1970). Behavioral and electrophysiological responses of male rats to female rat urine odors. In C. Pfaffmann (Ed.), *Olfaction and Taste III* (pp. 258–267). Rockefeller University Press: New York.

Pfaff, D. W. and Gregory, E. (1971). Olfactory coding in olfactory bulb and medial forebrain bundle on normal and castrated male rats. *Journal of Neurophysiology*, **34,** 208–216.

Price, J. L. (1987). The central olfactory and accessory olfactory systems. In T. E. Finger and

W. L. Silver (Eds.), *Neurobiology of Taste and Smell* (pp. 179–203). J. Wiley and Sons: New York.

Royet, J. P. (1983). Les aspects comportementaux de l'aversion conditionnée et de la néophobie. *Annales de Biologie,* **23,** 113–167.

Royet, J. P. and Pager, J. (1980). Réduction progressive de la néophobie au cours de l'expérience olfacto-gustative de la variété alimentaire chez le rat. *Comptes Rendus de l'Académie des Sciences (Paris),* **290,** 907–909.

Royet, J. P. and Pager, J. (1982). Lesions in olfactory pathways affecting neophobia and learned aversion differentially. *Behavioural Brain Research,* **4,** 251–262.

Royet, J. P., Sicard, G., Souchier, C. and Jourdan, F. (1987). Specificity of spatial patterns of glomerular activation in the mouse olfactory bulb: computer-assisted image analysis of 2-DG autoradiograms. *Brain Research,* **417,** 1–11.

Saucier, D. and Astic, L. (1986). Analysis of the topographical organization of olfactory epithelium projections in the rat. *Brain Research Bulletin,* **16,** 455–462.

Schaal, B. (1988). Olfaction in infants and children: developmental and functional perspectives. *Chemical Senses,* **13,** 145–190.

Scott, J. W. and Harrison, T. E. (1987). The olfactory bulb: anatomy and physiology. In T. E. Finger and W. L. Silver (Eds.), *Neurobiology of Taste and Smell* (pp. 151–178). J. Wiley and Sons: New York.

Skarda, C. and Freeman, W. J. (1987). How brains make chaos in order to make sense of the world. *Behavioral and Brain Sciences,* **10,** 161–195.

Slotnick, B. M., Graham, S., Laing, D. G. and Bell, G. A. (1987). Detection of proprionic acid vapor by rats with lesions of the olfactory bulb areas associated by the 2-deoxyglucose method. *Journal of Comparative Neurology,* **185,** 715–734.

Smithson, K. G., Weiss, M. L. and Hatton, G. I. (1989). Supraoptic nucleus afferents from the main olfactory bulb. I. Anatomical evidence from anterograde and retrograde tracer. *Neuroscience,* **31,** 277–288.

Smotherman, W. P. (1982). Odor aversion learning by the rat fetus. *Physiology and Behavior,* **29,** 769–771.

Smotherman, W. P. and Robinson, S. R. (1988). Behavior of rat fetuses following chemical or tactile stimulation. *Behavioral Neuroscience,* **102,** 24–34.

Staubli, U., Fraser, D., Faraday, R. L. and Lynch, G. (1987). Olfaction and the "data" memory system in rats. *Behavioral Neuroscience,* **101,** 757–765.

Steiner, J. E. (1979). Human facial expressions in response to taste and smell stimulation. *Advances in Study of Child Development & Behavior,* **13,** 257–295.

Stickford, G., Kimble, D. P. and Smotherman, W. P. (1982). *In utero* taste odor aversion conditioning of the rat. *Physiology and Behavior,* **28,** 5–7.

Sullivan, R. M., Wilson, P. A. and Leon, M. (1989). Norepinephrine and learning-induced plasticity in infant rat olfactory system. *Journal of Neuroscience,* **9,** 3998–4006.

Swindale, N. V. (1990). Is the cerebral cortex modular? *Trends in Neuroscience,* **13,** 487–492.

Takagi, S. F. (1984). The olfactory system of the Old World monkey. *Japanese Journal of Physiology,* **34,** 561–573.

Weiner, S. I., Paul, C. A. and Eichenbaum, H. (1989). Spatial and behavioral correlates of hippocampal neuronal activity. *Journal of Neuroscience,* **9,** 2737–2763.

Wysocki, C. J. and Meredith, M. (1987). The vomeronasal system. In T. E. Finger and W. L. Silver (Eds.), *Neurobiology of Taste and Smell* (pp. 125–150). J. Wiley and Sons: New York.

9

The neural control of feeding in primates

EDMUND T. ROLLS

Oxford University, Department of Experimental Psychology, South Parks Road, Oxford, England

DIRECT evidence on the neural processing involved in ingestion has been obtained from recordings of the activity of single neurones in several crucial regions of the brain in non-human primates and other mammals. These regions include systems that perform sensory analysis involved in the control of feeding, such as pathways, taste and olfaction in learning about foods including the amygdala and orbitofrontal cortex, and in the initiation of feeding behaviour such as the striatum. Some of the brain regions and pathways described in the text are shown in Fig. 9.1 on a lateral view of the brain of the macaque monkey.

9.1. Neuronal activity in the lateral hypothalamus during feeding

There is a population of neurones in the lateral hypothalamus and substantia innominata of the macaque with responses which are related to feeding (see Rolls, 1981a,b, 1986). In one sample of 764 hypothalamic neurones, 13.6% responded to the taste and/or sight of food. The neurones that responded to taste responded only when certain substances were in the mouth, such as glucose solution but not water or saline, had firing rates related to the concentration of the substance to which they responded (Rolls, Burton and Mora, 1980) and did not respond simply in relation to mouth movements (4.3% of the sample). The responses of the neurones associated with the sight of food occurred as soon as the monkey saw the food, before the food was in his mouth, and occurred only to foods and not to non-food objects. These neurones comprised 11.8% of one sample of 764 neurones (Rolls, Burton and Mora, 1976, 1980). Some of these neurones (2.5% of the total sample) responded to both the sight and taste of food (Rolls, Burton and Mora, 1976, 1980). The finding that there are neurones in the

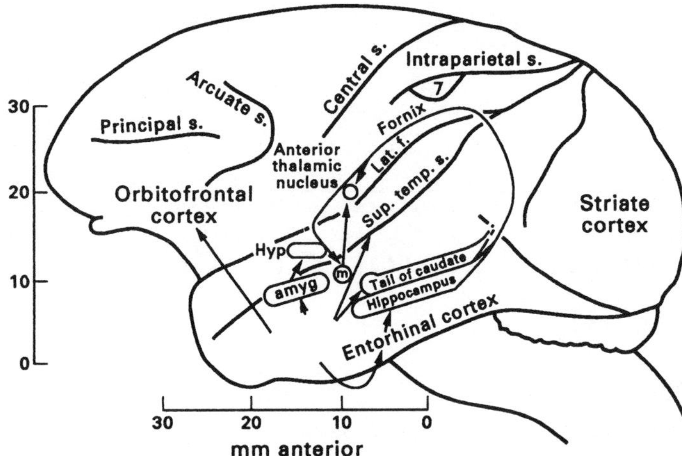

FIG. 9.1.   Some of the pathways described in the text are shown on this lateral view of the brain of the macaque monkey. amyg, amygdala; central s, central sulcus; Hyp, hypothalamus / substantia innominata / basal forebrain; Lat f, lateral (or Sylvian) fissure; m, mammillary body; Sup Temp s, superior temporal sulcus; 7, posterior parietal cortex, area 7.

lateral hypothalamus of the monkey which respond to the sight of food has been confirmed by Ono, Nishino, Sasaki, Fukuda and Muramoto (1980).

9.1.1. Neurones which respond to food

These are found in relatively small proportions in a region which includes the lateral hypothalamus and substantia innominata and extends from the lateral hypothalamus posteriorly through the anterior hypothalamus and lateral preoptic area to a region ventral to and anterior to the anterior commissure (see Fig. 9.3, reproduced from fig. 7 of Rolls *et al.*, 1979).

9.1.2. Effect of hunger

The responses of these neurones to the sight or taste of food only occur if the monkey is hungry (Burton, Rolls and Mora, 1976). Various signals of motivational state perform this modulation. Gastric distension is implicated by the finding that, after a monkey has fed to satiety, drainage of ingested food through a gastric cannula leads to the almost immediate resumption of feeding (Gibbs, Maddison and Rolls, 1981). (Because feeding is reinstated so rapidly, it is probably due to relief of distension rather than to the altered availability of chemical stimulation in the intestine). Another signal is provided by the presence of food in the duodenum and lower parts of the gut, as shown by the finding that normal satiety is not shown if ingested food is allowed to drain from a duodenal cannula (situated near the pylorus) and the monkey feeds almost continuously (Gibbs, Maddison and Rolls, 1981). Under these conditions, food does not accumulate normally in the stomach, showing that influences of duodenal or more distal origin are required to control gastric emptying, and thus to allow gastric distension to play a role in satiety. In this way an "enterogastric loop" contributes to satiety. The presence of food in the duodenum also contributes to satiety, as shown by the finding that duodenal infusions of food at rates similar to those of gastric emptying reduce the rate of feeding (Gibbs, Maddison and Rolls, 1981). Although there is no direct evidence in the monkey, it is very likely that rich information about nutrients in the gut and about gastric and intestinal distension reaches the nucleus of the solitary tract via the vagus, as it does in rats (Meï, 1993; Ewart, 1993). Other signals which influence hunger and satiety presumably reflect the metabolic state of the animal, and may include glucose level in the blood plasma (Le Magnen, 1992), perhaps sensed by cells in the hindbrain near the area postrema in the rat (Ritter, 1986). In the monkey, there are glucose-sensitive neurons in a number of hindbrain and hypothalamic sites, as shown by micro-electro-osmotic techniques (Oomura and Yoshimatsu, 1984; Aou *et al.*, 1984), but there is less evidence that these control food intake.

The hypothalamus is not necessarily the first stage at which processing of food-related stimuli is modulated by hunger. This issue is considered for the taste system below. Neuronal responses to visual stimuli in the inferior temporal cortex were not dependent on hunger (Rolls, Judge and Sanghera, 1977). The amygdala connects the inferior temporal visual cortex to the hypothalamus (see below) but the responses of an initial sample of neurones there were not found to depend on hunger (Sanghera, Rolls and Roper-Hall, 1979; Rolls, 1992). However, in the orbitofrontal cortex, which receives inputs from the inferior temporal visual cortex, and projects into the hypothalamus (see below and Russchen, Amaral and Price, 1985), neurones with visual responses to food are found, and neuronal responses to food in this region are modulated by hunger (Thorpe, Rolls and Maddison, 1983; see below). Thus, for visual processing, hunger modulates neuronal responsiveness only at late stages of sensory processing and thereafter in the hypothalamus. The adaptive value of modulation late in sensory processing is discussed below in connection with food-related taste processing where it also occurs.

9.1.3. Food-specific modulation of the responsiveness of lateral hypothalamic neurones and of appetite

During these experiments on satiety it was observed that, if a lateral hypothalamic neurone had ceased to respond to a food on which the monkey had been fed to satiety, then the neurone might still respond to a different food (see example in Fig. 9.2). This occurred for neurones with responses associated with the taste (Rolls, 1981b; Rolls *et al.*, 1986) or sight (Rolls and Rolls, 1982; Rolls *et al.*, 1986) of food. Corresponding to this neuronal specificity to sensory modality of

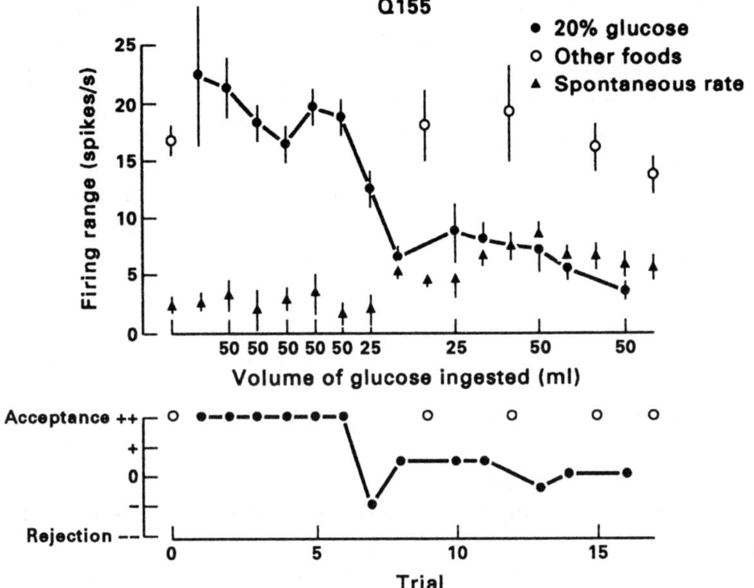

FIG. 9.2.   The effect of feeding the monkey to satiety with 20% glucose solution on the responses of a hypothalamic neurone to the taste of the glucose (filled circles) and to the taste of other foods (open circles). After the monkey had fed to satiety with glucose, the neurone responded much less to the taste of glucose, but still responded to the other foods. The satiety of the monkey, shown below, was measured by whether he accepted or rejected the glucose. (From Rolls *et al.*, 1986.)

the effects of feeding on satiety, the monkey rejected the food on which he had been fed to satiety, but accepted foods having other sensory characteristics on which he had not been fed.

As a result of these neurophysiological and behavioural observations showing the specificity of satiety in the monkey, experiments were performed to determine whether satiety was specific to foods just eaten in human subjects. It was found that the pleasantness of the taste of food eaten to satiety decreased more than for foods that had not been eaten (Rolls, Rolls, Rowe and Sweeney, 1981). One implication of this finding is that if one food is eaten to satiety, appetite reduction for other foods is often incomplete. This should mean that at least some of the other foods will be eaten, as seen in the monkey. This has been confirmed in an experiment in which either sausages or cheese with crackers were eaten for lunch. The rated liking for the food eaten decreased more than that for the food not eaten and, when an unexpected second course was offered, more was eaten if a subject had not been given that food in the first course than if he had been given that food in the first course (98% vs 40% of first-course intake eaten in the second course; Rolls, Rolls, Rowe and Sweeney, 1981). A further implication of these findings is that if a variety of foods is available, the total amount consumed will be more than when only one food is offered repeatedly. This prediction has been confirmed in a study in which people ate more when offered a variety of sandwich fillings than one filling or a variety of types of yoghurt which differed in taste, texture and colour (Rolls, Rowe, Rolls, Kingston, Megson and Gunary, 1981). It has also been confirmed in a study in which humans were offered a relatively normal meal of four courses: the change of food at each course significantly enhanced intake (Rolls, Van Duijenvoorde and Rolls, 1984).

Because sensory factors such as similarity of colour, shape, flavour and texture are usually more important than metabolic equivalence in terms of protein, carbohydrate and fat content in influencing how foods interact in this type of satiety, it has been termed "sensory-specific satiety" (Rolls and Rolls, 1977, 1982; Rolls, Rowe and Rolls, 1982; Rolls, Rolls, Rowe and Sweeney, 1981; Rolls, Rowe, Rolls, Kingston, Megson and Gunary, 1981; B. J. Rolls, 1990). As we shall see, this does not imply that a change in taste sensation is involved necessarily. It should be noted also that this effect is distinct from alliesthesia, which is defined as a change in the pleasantness of sensory inputs produced by internal signals such as glucose in the gut (Cabanac, 1971; Cabanac and Fantino, 1977; Cabanac and Duclaux, 1970). Sensory-specific satiety is defined as a change in the pleasantness of sensory inputs which is accounted for at least partly by the external sensory stimulation received (such as the taste of a particular food), as shown by its specificity to the food recently tasted. This also distinguishes it from conditioned food-specific satiety, which does not depend on recent prior presentation of the less pleasant food (Booth and Toase, 1983; Gibson and Booth, 1989).

The parallel between these studies of eating in people and of the neurophysiology of hypothalamic neurones in the monkey has been extended by the observations that, in humans, sensory-specific satiety occurs for the sight as well as for the taste of food (Rolls, Rowe and Rolls, 1982). Further, to complement the finding that in the hypothalamus neurones are found which respond differently to food and to water (Rolls and colleagues, unpublished observations), and that satiety with water can decrease the responsiveness of hypothalamic neurones which respond to water, satiety with water in man decreases the pleasantness of the sight and taste of water but not of food (Rolls, Rolls and Rowe, 1983).

The enhanced eating when a variety of foods is available, as a result of the operation of sensory-specific satiety, may have been advantageous in evolution in ensuring that different foods with important different nutrients were consumed. Today in human society, however, when a wide variety of foods is readily available, it may be a factor which can lead to overeating and obesity. Variety in food can lead to obesity in the rat (Rolls, Van Duijenvoorde and Rowe, 1983; see further B. J. Rolls and Hetherington, 1989).

In addition to the sensory-specific satiety described above which operates primarily within (see above) and in the post-meal period (Rolls, Van Duijenvoorde and Rolls, 1984), there is now evidence for a long-term form of sensory-specific satiety (Rolls and de Waal, 1985). People who had been in an Ethiopian refugee camp for 6 months rated their three regular foods to be less pleasant than three comparable foods which they had not been eating. The effect was long term in that it was not found in refugees who had been in the camp and eaten the regular foods for only 2 days (Rolls and de Waal, 1985). It may enhance nutrition under these circumstances to attempt to minimize the effect by providing some variety, perhaps even with spices (Rolls and de Waal, 1985).

9.1.4. Effects of learning

These responses of hypothalamic neurones in the primate become associated with the sight of food as a result of learning. The neurones come to respond to the sight of a previously neutral stimulus, such as a syringe, when the monkey is fed orally from it. The neurones cease to respond to a stimulus if it is no longer associated with food (in extinction or passive avoidance). Furthermore, the responses remain associated with whichever visual stimulus is associated with food in a visual discrimination task and its reversals (Mora, Rolls and Burton, 1976; Wilson and Rolls, 1990). This type of learning is important for it allows organisms to respond appropriately to environmental stimuli which previous experience has shown are foods. The brain mechanisms for this type of learning are discussed below.

The responses of these neurones suggest that they are involved in responses to food. Further evidence for this is that the responses of these neurones occur with relatively short latencies of 150–200 ms, and thus precede and predict the responses of the hungry monkey to food (Rolls, Sanghera and Roper-Hall, 1979).

9.1.5. Relation of neurone responses to the reward value of food

These hypothalamic neurones respond to food when it is rewarding, that is when the animal is not satiated and will work to obtain food. It is possible therefore that their responses are related to the reward value which food has for the hungry animal. Evidence consistent with this comes from studies with electrical stimulation of the brain. Stimulation of some brain regions is rewarding in that animals including man will work for passage of electrical current at some sites in those areas (Olds, 1977; Rolls, 1975, 1976, 1979). At some sites, including those in the lateral hypothalamus, the electrical stimulation appears to produce reward which is equivalent to food for the hungry animal, in that the animal will work hard to obtain the stimulation if he is hungry, but will work much less for it if he has been satiated (Olds, 1977; Hoebel, 1969). There is even evidence that the reward at some sites can mimic food for a hungry animal and at other sites water for a thirsty animal, in that rats chose electrical stimulation at one hypothalamic site when hungry and at a different site when thirsty (Gallistel and Beagley, 1971). It was therefore very interesting to discover that some of the neurones normally activated by food when the monkey was hungry were also activated by brain-stimulation reward (Rolls, 1975, 1976; Rolls, Burton and Mora, 1980). Further, it was shown that self-stimulation occurred through the recording electrode if it was near a region where hypothalamic neurones had been recorded which responded to food, and that this self-stimulation was attenuated by feeding the monkey to satiety (Rolls, Burton and Mora, 1980). Brain-stimulation reward also occurs at some brain sites (e.g. the orbitofrontal cortex and amygdala) which show effects of food and send projections to hypothalamic neurones (see below).

Self-stimulation of some brain sites might therefore be understood as the animal seeking to activate the neurones which he normally seeks to activate by food when he is hungry. This and other evidence (see Rolls, 1975, 1982) indicates that feeding normally occurs in order to obtain the sensory input produced by food which is rewarding when the animal is hungry.

9.2. Functions of the lateral hypothalamus in feeding

Lesions of the lateral hypothalamus which produce aphagia also damage fibre pathways coursing nearby such as the dopaminergic nigro-striatal bundle (Stricker and Zigmond, 1976). Damage to these pathways outside the lateral hypothalamus could produce aphagia (Marshall, Richardson and Teitelbaum, 1974). Damage to cells in the lateral hypothalamus without damaging fibres of passage, using as neurotoxins ibotenic acid or N-methyl-D-aspartate (NMDA) (Winn, Tarbuck and Dunnett, 1984; Dunnett, Lane and Winn, 1985; Clark *et al.*, 1991), produces a lasting decrease in food intake and body weight, however, which is not associated with dopamine-depleting damage to pathways or with the akinesia and sensorimotor deficits which are produced by damage to the dopamine systems (Winn, Tarbuck and Dunnett 1984; Dunnett, Lane and Winn, 1985; Clark *et al.*, 1991). These rats with lateral hypothalamic cellular lesions do not respond normally to acute blockade of glucose metabolism (Clark *et al.*, 1991).

Further information about the functions of these neurones could be provided by evidence on their output connections. Some hypothalamic neurones project to brainstem autonomic regions such as the dorsal motor nucleus of the vagus (Saper *et al.*, 1976, 1979). If some of the hypothalamic neurones with feeding-related activity projected in this way, their functions are likely to include the generation of autonomic responses to the sight of food. Some hypothalamic neurones project to the substantia nigra (Nauta and Domesick, 1978), and some neurones in the lateral hypothalamus and basal magnocellular forebrain nuclei of Meynert project directly to the cerebral cortex (Kievit and Kuypers, 1975; Divac, 1975; Heimer and Alheid, 1990). If some of these were feeding-related neurones, then by such routes they could influence whether feeding is initiated.

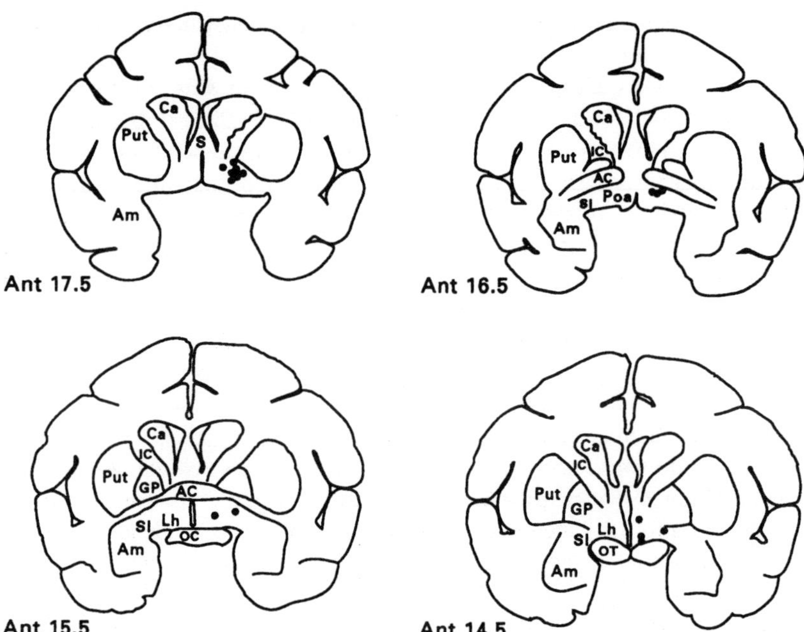

FIG. 9.3.   Sites in the lateral hypothalamus and basal forebrain of the macaque at which neurones were recorded that responded to the sight of food. *Abbreviations*: AC, anterior commissure; Am, amygdala; Ca, caudate nucleus; GP, globus pallidus; IC, internal capsule; Lh, lateral hypothalamus; OC, optic chiasm; OT, optic tract; Poa, preoptic area; Put, putamen; S, septal region; SI, substantia innominata. (From Rolls *et al.*, 1979.)

Electrical stimulation can be applied to these different regions, to determine whether hypothalamic neurones with feeding-related activity can be antidromically activated. It has so far been found in such experiments by E. T. Rolls, E. Murzi and C. Griffiths that some of these feeding-related neurones in the lateral hypothalamus and substantia innominata project directly to the cerebral cortex, to such areas as the prefrontal cortex in the sulcus principalis and the supplementary motor cortex. Thus at least some of these neurones project food-related information to the cerebral cortex, where it could be used in such processes as the initiation of ingestive behaviour. It also indicates that at least some of these neurons are in the basal magnocellular forebrain nuclei of Meynert, which is consistent with the reconstructions of the recording sites (Fig. 9.3). In addition, it seems quite likely that at least some of the feeding-related neurones influence brainstem autonomic motor neurones, because lesions of the lateral hypothalamus disrupt conditioned autonomic responses (LeDoux, Iwata, Cichetti and Reis, 1988).

There is thus an anatomical basis for the conclusion that the functions of the hypothalamus in feeding are related at least in part to the inputs which it receives from the forebrain conveying visual information about food, influenced by learning. (Such learned pattern-specific visual responses require forebrain areas such as the inferior temporal visual cortex and the amygdala, as described below). The hypothalamus and substantia innominata indeed receive projections from limbic structures such as the amygdala which in turn receive projections from the association cortex (Nauta, 1961; Herzog and Van Hoesen, 1976). The conclusion is also consistent with the evidence that decerebrate rats retain simple controls of feeding but do not show normal learning about foods (Grill and Norgren, 1978). These rats accept sweet solutions placed in their mouths when hungry and reject them when satiated, so that some control of responses to gustatory stimuli which depends on hunger can occur caudal to the level of the hypothalamus. However, these rats are unable to feed themselves and do not learn to avoid poisoned solutions.

Visual inputs and learning are important because animals, especially primates, may eat many foods every day, and must be able to select foods from other materials, as well as produce appropriate preparative autonomic responses such as salivation and the release of insulin. They must also be able to initiate appropriate actions in the environment to obtain food. Before any activation of motor neurones such as those that innervate the masticatory muscles, it is normally necessary to select which reinforcer in the environment should be the object of action, and then to select an appropriate (arbitrary) action to obtain the selected reinforcer. This indicates that direct connections to motor neurones from food recognition and reward systems are likely to be involved in the control of behaviour at only the lowest level in the sense of Hughlings Jackson (Swash, 1989). That is, food reward systems may be expected to project to an action control system, so that connections from the lateral hypothalamus, amygdala and orbitofrontal cortex to systems such as the basal ganglia are likely to be important as routes for the initiation of normal feeding (see section on the striatum below).

9.3. Activity in gustatory pathways during feeding

Since hunger modulates taste-responsive neurones in the hypothalamus, it may be asked whether this is because these cells are specially involved in the control of ingestion, or whether this type of modulation is evident throughout gustatory systems. It would be inefficient for motivational modulation to occur peripherally, because sensory information would be discarded without the possibility of central processing. A subjective parallel to such a situation would be if it were impossible to taste food (or even to see food) when satiated! It is perhaps more efficient for most of the sensory system to function similarly whether hungry or satiated, and to have a special system (such as the hypothalamus) following sensory processing where motivational state influences responsiveness.

Evidence on visual processing in primates in relation to food and hunger has been summarized above. In contrast, apparently there is at least some peripheral

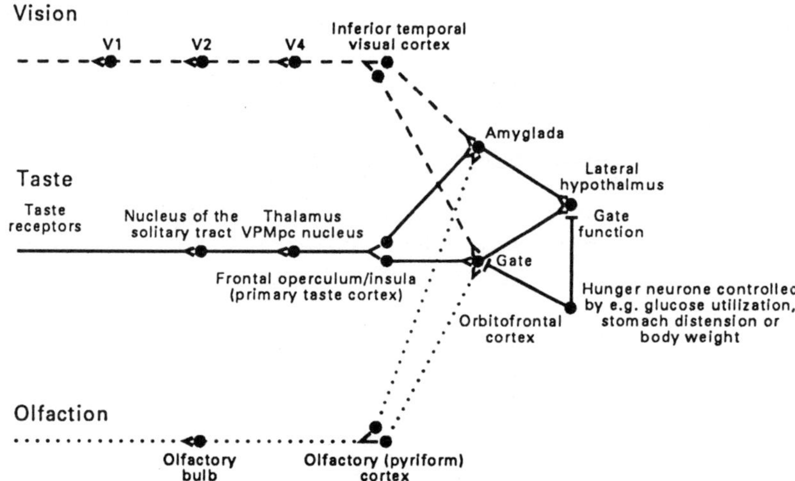

FIG. 9.4.   Schematic diagram showing some of the gustatory, olfactory, and visual pathways involved in processing sensory stimuli involved in the control of food intake.

modulation of taste processing in rats in the nucleus of the solitary tract (Scott and Giza, 1992). Evidence is now being obtained for primates on the tuning of neurones in the gustatory pathways, and on whether responsiveness at different stages is influenced by motivation, as follows. These investigations have also been able to show where flavour, that is a combination of gustatory and olfactory input, is computed in the primate brain. The gustatory and olfactory pathways, and some of their onward connections, are shown in Fig. 9.4.

The first central synapse of the gustatory system is in the rostral part of the nucleus of the solitary tract (Beckstead and Norgren, 1979; Beckstead, Morse and Norgren, 1980). In primates, neurones in this nucleus are relatively broadly tuned to the prototypical taste stimuli (sweet, salt, bitter and sour) (Scott, Yaxley, Sienkiewicz and Rolls, 1986a). The caudal half of this nucleus receives visceral afferents, and it is a possibility that such visceral information, for example about gastric distension, is used to modulate gustatory processing even at this early stage of the gustatory system. However, neuronal responses in the nucleus of the solitary tract to the taste of food are not influenced by whether the monkey is hungry or satiated by its own feeding (Yaxley, Rolls, Sienkiewicz and Scott, 1985).

The activity of single neurones in the primary gustatory cortex (frontal operculum and insula) during feeding in the monkey, is more sharply tuned to gustatory stimuli than activity in the nucleus of the solitary tract, with some neurones responding for example primarily to sweet and much less to salt, bitter or sour stimuli (Scott, Yaxley, Sienkiewicz and Rolls, 1986b; Yaxley, Rolls and Sienkiewicz, 1990). However, here also, hunger does not influence the

OFC

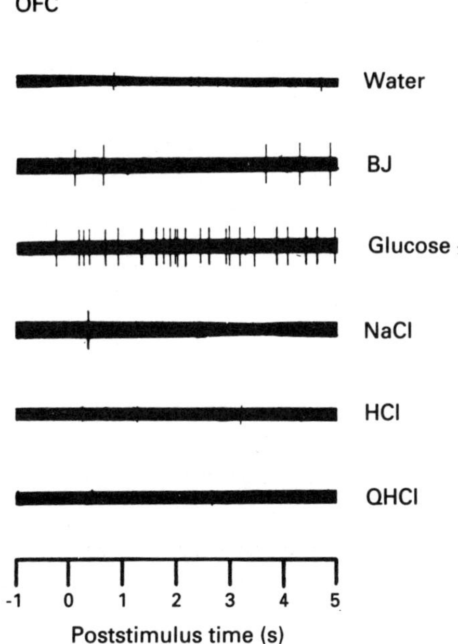

FIG. 9.5.   Examples of the responses recorded from one caudolateral orbitofrontal taste cortex neuron to the six taste stimuli, water, 20% blackcurrant juice (BJ), 1 M glucose, 1 M NaCl, 0. 01 M HCl, and 0. 001 M quinine HCl (QHCl). (From Rolls *et al.*, 1990)

magnitude of neuronal responses to gustatory stimuli (Rolls, Scott, Yaxley and Sienkiewicz, 1986; Yaxley, Rolls and Sienkiewicz, 1988).

A secondary cortical taste area, in the caudolateral orbitofrontal taste cortex of the primate, has recently been discovered. In this region, gustatory neurones are even more sharply tuned to particular taste stimuli (Rolls, Yaxley and Sienkiewicz, 1990; Rolls and Treves, 1990) (see Fig. 9.5). In addition to representations of sweet, salt, bitter and sour stimuli, different neurones in this region respond to compounds such as glutamate (Baylis and Rolls, 1991) and to a wide range of complex foods (Baylis and Rolls, in preparation). In this region, the responses of taste neurones to a particular food decrease to zero when a monkey is fed on it to satiety (Rolls, Sienkiewicz and Yaxley, 1989). That is, this modulation is sensory-specific (see, e.g., Fig. 9.6).

After eating to satiety, people rated the taste of the food on which they had been satiated to be almost as intense as it was when they were hungry, though much less pleasant (Rolls, Rolls and Rowe, 1983). This parallels the insensitivity to satiety of activity in the frontal opercular and insular taste cortices as well as the nucleus of the solitary tract. On the other hand, the responses of the neurones in the orbitofrontal taste area and in the lateral hypothalamus are modulated by satiety. It may be in areas such as these that neuronal activity is

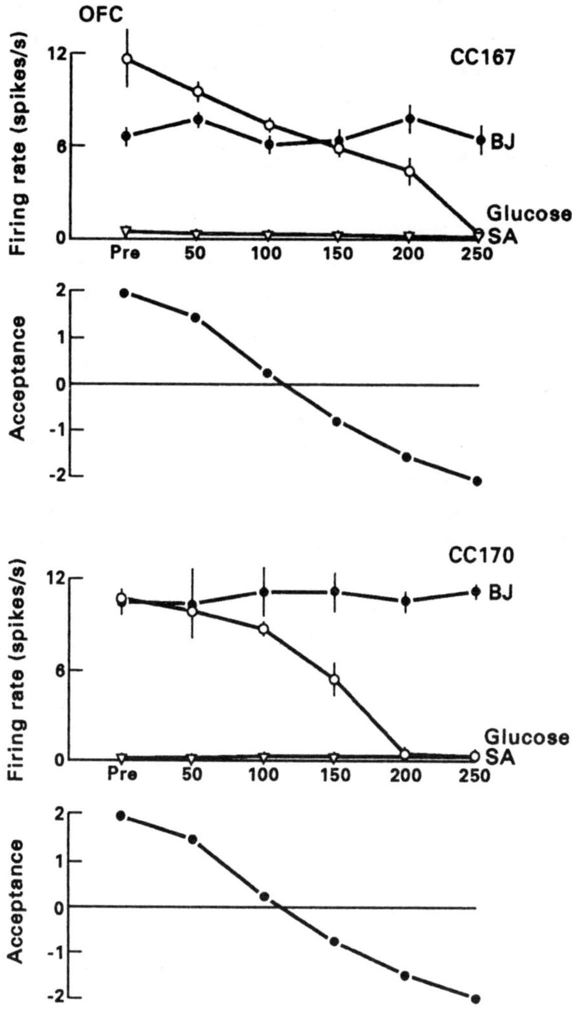

FIG. 9.6.   The effect of feeding to satiety with glucose solution on the responses of two neurones in the secondary taste cortex to the taste of glucose and of blackcurrant juice (BJ). The spontaneous firing rate is also indicated (SA). Below the neuronal response data for each experiment, the behavioural measure of the acceptance or rejection of the solution on a scale from $+2$ to -2 (see text) is shown. The solution used to feed to satiety was 20% glucose. The monkey was fed 50 ml of the solution at each stage of the experiment as indicated along the abscissa, until he was satiated as shown by whether he accepted or rejected the solution. Pre: the firing rate of the neurone before the satiety experiment started. (From Rolls, Sienkiewicz and Yaxley, 1989.)

related to whether a food tastes pleasant, and is rewarding, that is it reinforces work to obtain the food. Sensory-specific satiety cannot therefore be accounted for by adaptation at the receptor level, in the nucleus of the solitary tract, in the frontal opercular or insular gustatory cortices, i.e., in the primary gustatory cortex; otherwise modulation of neuronal responsiveness during feeding should have been apparent in the recordings made in these regions.

These findings lead to the following proposal of a neuronal mechanism for sensory-specific satiety (see also Rolls and Treves, 1990). The tuning of neurones becomes more specific for gustatory stimuli from the nucleus of the solitary tract, through the gustatory thalamus and on to frontal opercular taste cortex. Satiety, habituation and adaptation are not features of responses in these regions. The tuning of neurones becomes even more specific in the orbitofrontal cortex, but here there is some effect of satiety by internal signals such as gastric distension and glucose utilization, and in addition habituation sets in over several minutes and lasts for 1–2 hours. Because of the relative specificity of the tuning of orbitofrontal taste neurones, this results in a decrease in the response to that food, but different foods continue to activate other neurones. (For orbitofrontal cortex neurones that respond to two similar tastes before satiety, it is suggested that the habituation that results in a loss of the response to the taste eaten to satiety occurs because of habituation of the afferent neurones or synapses onto these orbitofrontal cortex neurones.) Other parts of the brain to use the activity of the orbitofrontal cortex neurones to decide the reward value of that particular taste. Evidence that the activity of neurones in the orbitofrontal cortex does reflect reward includes the observation that electrical stimulation here produces reward which is attenuated by satiety (Mora, Avrith, Phillips and Rolls, 1979). One output of these neurones may be to the hypothalamic neurones with food-related responses which decrease specifically to a food which has just been eaten to satiety. Another output may be to the ventral and adjoining striatum, which may provide an important link between reward systems and action (see below).

It is suggested that the computational significance of this architecture is as follows (see also Rolls, 1986, 1989; Rolls and Treves, 1990). If satiety were to operate at an early level of sensory analysis, where neurones are broadly tuned, responses to non-foods would become attenuated as well as responses to foods; this could be dangerous if poisonous non-foods became undetectable. Also, unless tuning were relatively fine, reduction in neuronal firing after one food had been eaten would inevitably reduce behavioural responsiveness to other foods not eaten in the meal which in fact remain relatively pleasant. A sensory-specific satiety mechanism can be built by arranging for tuning to particular foods to become relatively specific at one level of the nervous system (as a result of processing in earlier stages), and then at this stage (but not at prior stages) to allow habituation to be a property of the synapses, as proposed above.

Thus information processing in the taste system illustrates an important principle of higher nervous system function in primates, namely that it is only after several or many stages of sensory information processing (which produce efficient categorization of the stimulus) that there is an interface to motivational systems, to other modalities, or to systems involved in association memory (Rolls, 1987; Rolls and Treves, 1990).

9.3.1. Convergence between gustatory and olfactory processing to represent flavour

At some stage, taste representations are brought together with inputs from different modalities. For example, gustatory and olfactory inputs together are needed to form a representation of flavour. Takagi and his colleagues (Tanabe *et al.*, 1975a,b) have found an olfactory area in the medial orbitofrontal cortex. In a mid-mediolateral part of the caudal orbitofrontal cortex is the area investigated by Thorpe, Rolls and Maddison (1983) in which are found many neurones with visual responses and some with gustatory responses. Our impression was that the caudolateral orbitofrontal taste cortex was different from the frontal opercular and insular primary taste cortices, in that there were neurones with responses in other modalities within or very close to the area. We therefore investigated systematically whether there are neurones in the secondary taste cortex which respond to stimuli in other modalities, including the olfactory and visual modalities, and whether single neurones in this cortical region in some cases respond to stimuli from more than one modality (Rolls, 1989; Baylis and Rolls, in preparation).

We found that, of the single neurones which responded to any of these modalities, many were unimodal (taste 47%, olfactory 12%, visual 10%) but in close proximity to each other. Some neurones showed convergence, responding for example to gustatory and visual inputs (17%), gustatory and olfactory inputs (10%), and olfactory and visual inputs (4%). Some of these multimodal neurones had functionally corresponding sensitivities in the two modalities. For example, some single neurones responded best to sweet tastes (e.g. 1 M glucose) and responded more in a visual discrimination task to the visual stimulus which signified sweet fruit juice than to that which signified saline. Others responded to sweet taste, as well as to fruit odour in an olfactory discrimination task. An example of one such bimodal neurone is shown in Fig. 9.7. The neurone responded best among the tastants to NaCl (N), and best among the odours to onion odour (On), and well also to salmon (S). The olfactory input to these neurones was further defined by measuring their responses while the monkey performed an olfactory discrimination task, in which it was possible to show that these neurones could respond to odours with response latencies as short as 150 ms, and had selective responses in that the neurones in some cases did not respond while the monkey performed a visual discrimination task (see Rolls, 1989; Fig. 9.9). The different types of neurones (unimodal in different modalities, and multimodal) were frequently found close to one another in tracks made into this region (see Fig. 9.8), consistently with the hypothesis that the multimo-

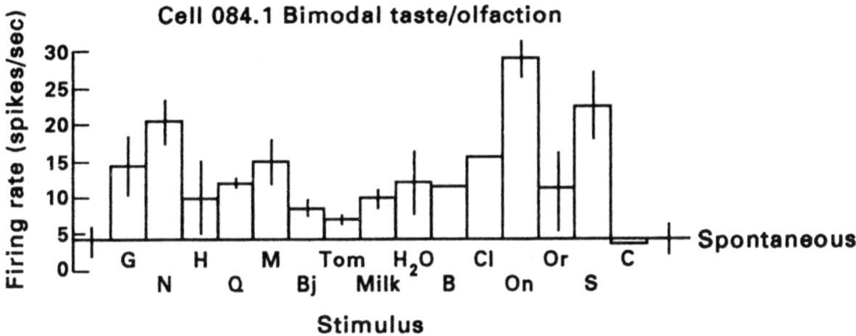

FIG. 9.7. The responses of a bimodal neurone recorded in the caudolateral orbitofrontal cortex. G,—1 M glucose; N, 0.1 M NaCl; H, 0.01 M HCl; Q, 0.001 M Quinine HCl; M, 0.1 M monosodium glutamate; Bj, 20% blackcurrant juice; Tom, tomato juice; B, banana odour; Cl, clove oil odour; On, onion odour; Or, orange odour; S, salmon odour; C, control no-odour presentation. The mean responses ± se are shown. The neurone responded best to the tastes of NaCl and monosodium glutamate and to the odours of onion and salmon.

dal representations are actually being formed from unimodal inputs to this region (see also Rolls and Mason, 1991). In current experiments, we (E. T. Rolls and H. Critchley) are finding that the responses of these olfactory neurones decrease selectively for the odour of a food with which the monkey is fed to satiety. Thus the activity of these neurones is related to sensory-specific satiety effects for odour.

These results show that there are regions in the orbitofrontal cortex of primates where the sensory modalities of taste and olfaction converge, often in neurones having sensitivities across the two modalities that correspond to a food and showing food-specific satiety. This may be the part of the primate nervous system where food flavour is computed and habituation to the flavour of a food being eaten occurs.

9.4. Activity in amygdala and temporal cortex related to feeding

Bilateral damage to the temporal lobes of primates leads to the Kluver-Bucy syndrome, in which monkeys for example select non-food as well as food items and place them in their mouths and repeatedly fail to avoid noxious stimuli (Kluver and Bucy, 1939; Jones and Mishkin, 1972; Aggleton and Passingham, 1982; Baylis and Gaffan, 1991). Rats with lesions in the basolateral amygdala also display altered food selection, in that they ingest relatively novel foods (Rolls, E. T. and Rolls, B. J., 1973; Borsini and Rolls, 1984) and do not learn to avoid ingesting a solution which has previously resulted in sickness (Rolls, B. J. and Rolls, E. T., 1973). (The deficit in learned taste avoidance in rats may be due to damage to the insular taste cortex, which has projections through and to the amygdala—Dunn and Everitt, 1988.) The basis for these alterations in food

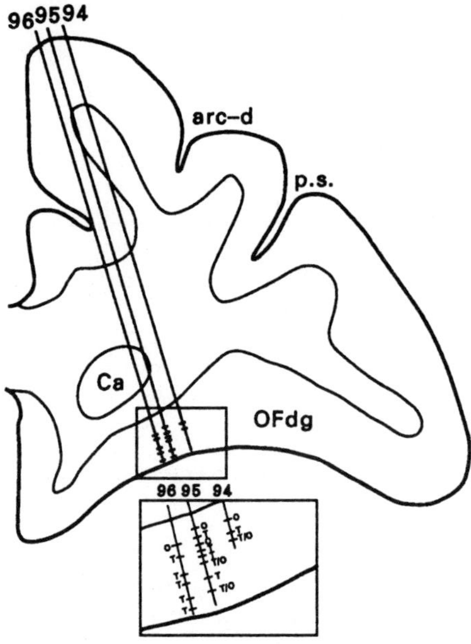

Fig. 9.8.   Examples of tracks made into the orbitofrontal cortex in which taste (T) and olfactory (O) neurones were recorded close to each other in the same tracks. Some of the neurones were bimodal (T/O).

selection and in food-related learning are considered next (see also Rolls, 1990b, 1992b).

The monkeys with temporal lobe damage have a visual discrimination deficit, in that they are impaired in learning to select one of two objects under which food is found, and thus fail to form correctly an association between the visual stimulus and reinforcement (Jones and Mishkin, 1972; Gaffan, 1992). Gaffan and Harrison (1987) and Gaffan, Gaffan and Harrison (1988) have shown that the tasks which are impaired by amygdala lesions in monkeys typically involve a cross-modal association from a previously neutral stimulus to a primary reinforcing stimulus (such as the taste of food), consistent with the hypothesis that the amygdala is involved in learning associations between stimuli and primary reinforcers (see also Gaffan, 1992; Gaffan, Gaffan and Harrison, 1989). Further evidence linking the amygdala to reinforcement mechanisms is that monkeys will work in order to obtain electrical stimulation of the amygdala, and that single neurones in the amygdala are activated by brain-stimulation reward of a number of different sites (Rolls, 1975; Rolls, Burton and Mora, 1980).

The Kluver-Bucy syndrome is produced by lesions which damage the cortical areas in the anterior part of the temporal lobe and the underlying amygdala (Jones and Mishkin, 1972) or by lesions of the amygdala (Weiskrantz, 1956; Aggleton and Passingham, 1981; Gaffan, 1992) or of the temporal lobe of

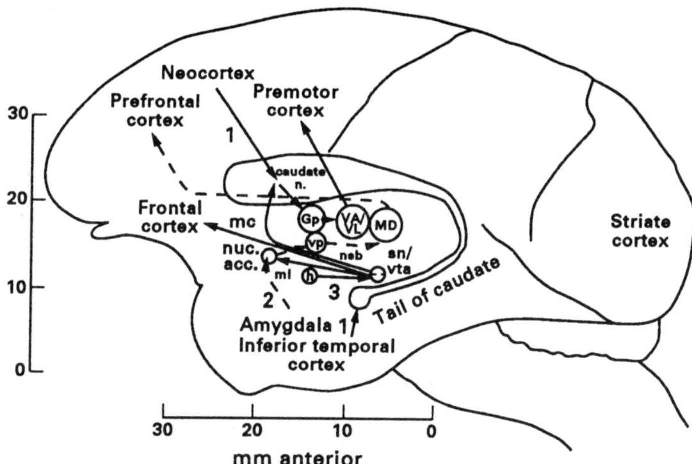

FIG. 9.9.   Some of the striatal and connected regions in which the activity of single neurones is described shown on a lateral view of the brain of the macaque monkey. Gp, globus pallidus; h, hypothalamus; Sn, substantia nigra, pars compacta (A9 cell group), which gives rise to the nigrostriatal dopaminergic pathway, or nigrostriatal bundle (nsb); vta, ventral tegmental, area, containing the A10 cell group, which gives rise to the mesocortical dopamine pathway (mc) projecting to the frontal and cingulate cortices and to the mesolimbic dopamine pathway (ml), which projects to the nucleus accumbens (nuc acc).

neocortex (Akert *et al.*, 1961). Lesions to part of the temporal lobe, damaging the inferior temporal visual cortex and extending into the cortex in the ventral bank of the superior temporal sulcus, produce visual aspects of the syndrome, seen for example as a tendency to select non-food as well as food items (Weiskrantz and Saunders, 1984). There are connections from the inferior temporal visual cortex to the amygdala (Herzog and Van Hoesen, 1976), which in turn projects to the hypothalamus (Nauta, 1961), thus providing a route for visual information to reach the hypothalamus (see Rolls, 1981b, 1990b, 1992b; Amaral, 1992). Since damage to hypothalamic cells can disrupt feeding, there may be a system which projects from visual cortex in the temporal lobe to the amygdala, and further connects to structures such as the lateral hypothalamus, which is involved in behavioural responses made on the basis of learned associations between visual stimuli and primary (unlearned) reinforcers such as the taste of food (see Fig. 9.4). The contribution of each of these regions to the visual analysis and learning required for food selection can be further elucidated by evidence   from   the activity of single neurones.

Recordings were made in the inferior temporal visual cortex while rhesus monkeys performed visual discriminations, and while they were shown visual stimuli associated with positive reinforcement such as food, with negative reinforcement such as aversive hypertonic saline, and without reinforcement (Rolls,

Judge and Sanghera, 1977). During visual discriminations, inferior temporal neurones often had sustained visual responses with latencies of 100–140 ms to the discriminanda, but these responses did not depend on whether the visual stimuli were associated with reward or punishment. This observation is consistent with the findings of Ridley, Hester and Ettlinger (1977), Jarvis and Mishkin (1977), Gross, Bender and Gerstein (1979), and Sato *et al.* (1980). Further, it was found that inferior temporal neurones did not respond only to food-related visual stimuli, or only to aversive stimuli, and were not dependent on hunger, but that rather in many cases their responses depended on physical aspects of the stimuli such as shape, size, orientation, colour, or texture (Rolls, Judge and Sanghera, 1977).

Thus the responses of neurones in the inferior temporal visual cortex do not reflect the association of visual stimuli with reinforcers such as food. Given these findings and the lesion evidence described above, it is thus likely that the inferior temporal cortex is an input stage for this associative process.

On the basis of anatomical connections (see Fig. 9.4) the next structure activated is the amygdala. In recordings from 1754 amygdaloid neurones, 113 (6.4%) had visual responses which in most cases were sustained while the monkey looked at effective visual stimuli (Sanghera, Rolls, and Roper-Hall, 1979). The latency of the responses was 100–140 ms or more. Many were in a dorsolateral region of the amygdala known to receive inputs directly from the inferior temporal visual cortex (Herzog and Van Hoesen, 1976). The majority (85%) of these visual neurones responded more strongly to some stimuli than to others, but the responses could not usually be accounted for by physical factors such as orientation, colour or texture. It was found that 22 (19.5%) of these visual neurones responded primarily to foods and to objects associated with food, but for none of these neurones did the responses occur uniquely to food-related stimuli; they all responded also to one or more aversive or neutral stimuli. Further, although some neurones responded in a visual discrimination to the stimulus which indicated food reward but not to the stimulus associated with aversive saline, only minor modifications of the neuronal responses were obtained when the association of the stimuli with reinforcement was switched in reversal of the visual discrimination. Thus even the responses of these neurones were not invariably related to whichever stimulus was associated with reward (see further Rolls, 1992b). A comparable population of neurones with responses apparently partly but not uniquely related to aversive visual stimuli was also found (Sanghera, Rolls and Roper-Hall, 1979).

Amygdala neurones with responses which are probably similar to these have also been described by Ono *et al.* (1980), Nishijo *et al.* (1988), Ono *et al.* (1989), and Ono and Nishijo (1992). When Nishijo *et al.* (1988) tested four amygdala neurones in a simpler relearning situation than reversal in which salt was added to a piece of food such as a water melon, the neurones' responses to the sight of the water-melon appeared to diminish. However, in this task it was not clear whether the monkeys continued to look at the stimuli during extinction. It will be of interest in further studies to investigate whether evidence can be found in extinction for a rapid decrease in the neuronal responses to visual stimuli formerly associated with reward, even when fixation of the stimuli is adequate (see Rolls, 1992b).

Wilson and Rolls (1993; see also Rolls, 1992b) extended the analysis of the responses of these amygdala neurones by showing that, while they do respond to (some) stimuli associated with primary reinforcement, they do not respond if the reinforcement must be determined on the basis of a rule (such as stimuli when novel are negatively reinforced, and when familiar are positively reinforced). This is consistent with the evidence that the amygdala is involved when reward is based on simple association of a stimulus with a reinforcer such as the taste of food as normally occurs during feeding, but is not involved when reinforcement must be obtained in some other way (see Gaffan, 1992; Rolls, 1992b). In the same study (Wilson and Rolls, 1993), it was shown that these amygdala neurones that respond to food can also respond to some other stimuli while they are relatively novel. It is suggested that this mechanism is involved when relatively novel stimuli are encountered and they are investigated, e.g. by being smelled and then placed in the mouth, to assess whether they are foods (see Rolls, 1992b).

Thus, the amygdala could be involved at an early stage of the processing by which visual stimuli are associated with reinforcement, but neurones with responses more closely related to reinforcement are found in areas to which the amygdala projects, such as the lateral hypothalamus, substantia innominata, and ventral striatum. Through these stages, neuronal responses become more relevant to the control of feeding, so that finally there are neurones in, for example, the lateral hypothalamus and substantia innominata with responses which occur only to stimuli which the organism has learned are food or signify food and to which it is appropriate when hungry to initiate feeding (see Fig. 9.4).

9.5. Activity in orbitofrontal cortex related to feeding

Monkeys with damage to the orbitofrontal cortex select and eat foods which are normally rejected (Butter, McDonald and Snyder, 1969; Baylis and Gaffan, 1991). Their food choice behaviour is very similar to that of monkeys with amygdala lesions (Baylis and Gaffan, 1991). Lesions of the orbitofrontal cortex also lead to a failure to correct feeding responses when these become inappropriate. Examples of the situations in which these abnormalities in feeding responses are found include: (a) extinction, where feeding responses continue to be made to the previously reinforced stimulus; (b) reversals of visual discriminations, when the monkeys with lesions make responses to the previously reinforced stimulus or object; (c) go/no-go tasks, where responses are made to the stimulus which is not associated with food reward; and (d) passive avoidance, in which feeding responses are made even when they are punished (Butter, 1969; Iversen and Mishkin, 1970; Jones and Mishkin, 1972; Tanaka, 1973; see also Rosenkilde, 1979; Fuster, 1989). It may be noted, however, that the formation of associations between visual stimuli and reinforcement is less affected by these lesions than by temporal lobe lesions, as tested during visual discrimination learning and reversals (Jones and Mishkin, 1972).

To investigate how the orbitofrontal cortex may be involved in feeding and in the correction of feeding responses when these become inappropriate, recordings were made of the activity of 494 orbitofrontal neurones during the performance of a go/no-go task, reversals of visual discrimination, extinction and passive avoidance (Thorpe, Rolls and Maddison, 1983). First, neurones were found which responded in relation to the preparatory auditory or visual signal used before each trial (15.1%), or non-discriminatively during the period in which the discriminative visual stimuli were shown (37.8%). These neurones are not considered further here. A second group of neurones (8.6%) responded discriminatively during the period in which the visual stimuli were shown. The majority of these neurones responded to whichever visual stimulus was associated with reward, in that the stimulus to which they responded changed during reversal. However, six of these neurones required a combination of a particular visual stimulus in the discrimination *and* reward in order to respond. Furthermore, none of this second group of neurones responded to all the reward-related stimuli including different foods which were shown, so that in general this group of neurones coded for a combination of one or several visual stimuli *and* reward. Thus information that particular visual stimuli had previously been associated with reinforcement was represented in the responses of orbitofrontal neurones. A third group of neurones (9.7%) responded after the lick response to obtain reward had been made. Some of these responded whether fruit juice reward or aversive hypertonic saline was obtained on trials on which the monkey licked in error or saline was given in the first trials of a reversal. Hence it seems that information that a lick had been made was represented in the orbitofrontal cortex through these neurones. Other neurones in this third group responded only when fruit juice was obtained, and so these neurones represent the information that reward had been given on that trial. Such neurones reflect the taste of the liquid received, and are in a part of the orbitofrontal cortex which is close to, and probably receives inputs from, the secondary taste cortex (Rolls, 1989; Rolls *et al.*, 1990). Still other neurones in this group responded when saline was obtained after a response was made in error, when saline was obtained on the first few trials of a reversal (but not in either situation when saline was simply placed in the mouth), when reward was not given in extinction, or when food was taken away instead of being given to the monkey, but did not respond in all these situations when reinforcement was omitted or punishment was given. Thus task-selective information that reward had been omitted or punishment given was represented in the responses of these neurones.

These three groups of neurones found in the orbitofrontal cortex could together provide for computation of whether the reinforcement previously associated with a particular stimulus was still being obtained, and generation of a signal if a match was not obtained. This signal could be partly reflected in the responses of the last subset of neurones with task-selective responses to non-reward or to unexpected punishment. This signal could be used to alter the monkey's behaviour, leading for example to reversal to one particular stimulus but not to other stimuli, to extinction to one stimulus but not to others, etc. It could also lead to the altered responses of the orbitofrontal differential neurones found as a result of learning in reversal, so that their responses indicate appropriately whether a particular stimulus is now associated with reinforcement.

9.6. Functions of orbitofrontal cortex in feeding and reward

Thus the orbitofrontal cortex contains neurones which appear to be involved in altering behavioural responses when these are no longer associated with reward or become associated with punishment. In the context of feeding, it appears that without these neurones the primate is unable to suppress his behaviour correctly to non-food objects, in that altered food preferences are produced by orbitofrontal damage (Butter *et al.*, 1969). It also appears that without these neurones the primate is unable to correct his behaviour when it becomes appropriate to break a learned association between a stimulus and a reward such as food (Jones and Mishkin, 1972). The orbitofrontal neurones could be involved in the actual breaking of the association or in the alteration of behaviour when other neurones signal that the connection is no longer appropriate. As shown here, the orbitofrontal cortex contains neurones with responses which could provide the information necessary for, and the basis for, the unlearning. This type of unlearning is important in enabling animals to alter the environmental stimuli to which motivational responses such as feeding have previously been made, when experience shows that such responses have become inappropriate. In this way they can ensure that their feeding and other motivational responses remain continually adapted to a changing environment.

The more rapid reversal of neuronal responses in the orbitofrontal cortex, and in a region to which it projects, the basal forebrain (Thorpe, Rolls and Maddison, 1983; Wilson and Rolls, 1990), than in the amygdala suggest that the orbitofrontal cortex is more involved than the amygdala in the rapid readjustments of behavioural responses made to stimuli when their reinforcement value is repeatedly changing, as in discrimination reversal tasks (Thorpe *et al.*, 1983; Rolls, 1986, 1990b). The ability to alter responses flexibly to stimuli according to changes in their reinforcement associations is important in motivated behaviour such as ingestion and in emotional behaviour. It is suggested that the orbitofrontal cortex adds flexibility to a more basic capacity which the amygdala alone implements for stimulus-reinforcement learning (Rolls, 1986, 1990b). The orbitofrontal cortex is greatly developed in primates. Yet its connections are similar to those of the amygdala, and it is connected to the amygdala. It can be suggested therefore that as part of corticalization of functions in evolution, the orbitofrontal cortex has come to be hierarchically above the amygdala, having special importance when rapid readjustment of stimulus-reinforcement associations is required (Rolls, 1990b). This suggestion is also consistent with the indication that, whereas in rodents subcortical structures such as the amygdala and hypothalamus have access to taste information from the precortical taste system, cortical processing of taste is of great importance in primates. It is very appropriate that the orbitofrontal cortex area just described is found just medial to the secondary taste cortex in primates and receives inputs from the visual association cortex (inferior temporal cortex), the olfactory (pyriform) cortex, and probably from the somatosensory cortex, so that reward associations between these different modalities can be determined rapidly.

9.7. Processing in the striatum during feeding

Damage to the nigrostriatal bundle, which depletes the striatum of dopamine, produces aphagia and adipsia associated with a sensori-motor disturbance in the rat (Ungerstedt, 1971; Marshall, Richardson and Teitelbaum, 1974; Stricker and Zigmond, 1976; Stricker, 1984). Moreover, many of the brain systems implicated in the control of feeding, such as the amygdala and orbito-frontal cortex, have projections to the striatum, which could provide a route for these brain systems to lead to feeding responses (Mogenson *et al.*, 1980; Rolls and Williams, 1987a; Rolls and Johnstone, 1992; Williams *et al.*, 1993). In order to analyse how striatal function is involved in feeding, the activity of single neurones is recorded in different regions of the striatum (see Fig. 9.9) during the initiation of feeding and during the performance of other tasks known to be affected by damage to particular regions of the striatum (see Rolls, 1979, 1984, 1986; Rolls and Williams, 1987a; Rolls and Johnstone, 1992).

In the *head of the caudate nucleus* (Rolls, Thorpe and Maddison, 1983), which receives inputs particularly from the prefrontal cortex, many neurones responded to environmental stimuli which were cues to the monkey to prepare for the possible initiation of a feeding response. Thus, 22.4% of neurones recorded responded during a cue given by the experimenter that a food or non-food object was about to be shown to the monkey (and fed if food). In a visual discrimination task made to obtain food, 14.5% of the neurones (including some of the above) responded during a 0.5-sec tone/light cue which preceded and signalled the start of each trial. It is suggested that these neurones are involved in the utilization of environmental cues for the preparation for movement, and that disruption of the function of these neurones contributes to the akinesia or failure to initiate movements (including those required for feeding) found after depletion of dopamine in the striatum (Rolls, Thorpe and Maddison, 1983).

Some other neurones (25.8%) in the head of the caudate responded if food was shown to the monkey immediately prior to feeding, but not when food-related-visual stimuli were shown, such as during the visual discrimination task. Comparably, some other neurones (24.3%) responded differentially in the visual discrimination task, for example, to the visual stimulus which indicated that the monkey could initiate a lick response to obtain food. Yet these neurones typically did not respond when food was simply shown to the monkey prior to feeding. Thus these neurones respond to particular stimuli which indicate that particular motor responses should be made. They are situation -specific. Hence it is suggested that these neurones are involved in connections of stimuli to motor responses. In that their responses are situation-specific, they differ from the hypothalamic neurones that respond to the sight of food (Rolls, Thorpe and Maddison, 1983).

It is thus suggested that these neurones in the head of the caudate nucleus could be involved in relatively fixed feeding responses made in particular, probably well-learned, situations but do not provide a signal which reflects whether a visual stimulus is associated with food, on the basis of which a response required to obtain the food could be initiated. Rather, it is likely that the systems in the temporal lobe, hypothalamus and orbitofrontal cortex are involved in this more flexible decoding of the food value of visual stimuli.

In the *tail of the caudate nucleus*, which receives inputs from the inferior temporal visual cortex, neurones responded to visual stimuli such as gratings and edges and showed habituation which was rapid and pattern-specific (Caan, Perrett and Rolls, 1984). It was suggested that these neurones are involved in orientation to patterned visual stimuli and habituation to specific stimuli. These neurones would thus appear not to be involved directly in the control of feeding, although a disturbance in the ability to orient normally to a changed visual stimulus could have an indirect effect.

In the *putamen*, which receives connections from the sensori-motor cortex, neurons were found with activity related to movements made by the monkey for example with the mouth or arm (Rolls *et al*, 1984). Disturbances in the normal function of these neurones might be expected to affect the ability to initiate and execute movements, and thus also might indirectly affect the ability to feed normally.

The *ventral striatum*,which includes the nucleus accumbens, the olfactory tubercle (or anterior perforated substance of primates), and the islands of Calleja, receives inputs from limbic structures such as the amygdala and hippocampus, and also from the orbitofrontal cortex, and projects to the ventral pallidum (Groenewegen *et al.*, 1991). The ventral pallidum may then influence motor output by the subthalamic nucleus/globus pallidus/ventral thalamus/pre-motor cortex route or via the mediodorsal nucleus of the thalamus and prefrontal cortex (Heimer *et al.*, 1982). The ventral striatum may thus be for limbic structures what the neostriatum is for neocortical structures, that is a route from limbic to output regions. There is evidence linking the ventral striatum and its dopamine input to reward, for manipulations of this system alter the incentive

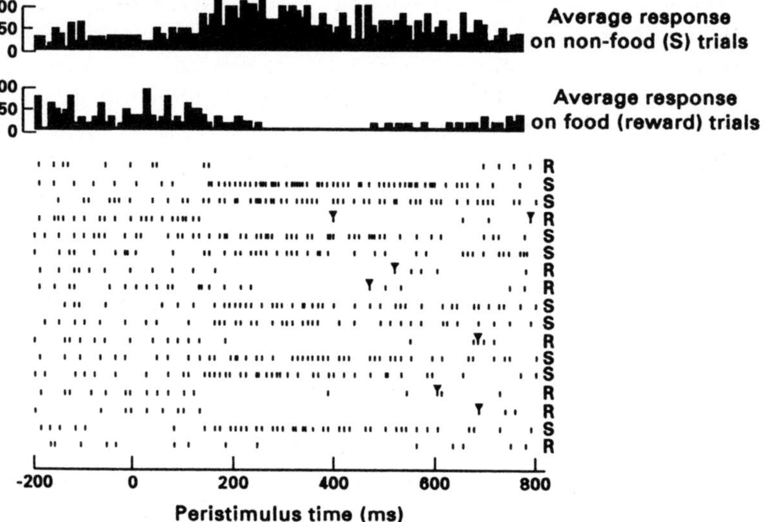

FIG. 9.10.   Responses of a neurone in the visual discrimination task. The neurone decreased its firing rate to the S + on food reward trials, and increased its firing rate to the S − on non-food trials. Rastergrams and peristimulus time histograms are shown.

effects which learned rewarding stimuli have on behaviour (Everitt and Robbins, 1992; Robbins and Everitt, 1992).

Neurones in the ventral striatum of the behaving monkey show a number of different types of response (Rolls and Williams, 1987b, 1987a; Williams *et al.*, 1993). One population of neurones responded to visual stimuli that are reinforcing (Rolls, 1990b). (Reinforcers are stimuli which, if their occurrence, termination or omission is made contingent upon the making of a behavioural response, alter the future emission of that response.) The responses of an example of a neurone of this type are shown in Fig. 9.10. The neurone increased its firing rate to the S − on non-food trials in the visual discrimination task, and decreased its firing rate to the S + on food reward trials. The differential response latency of this neurone to the reward-related and to the saline-related visual stimulus was approximately 150 ms (see Fig. 9.10), and this value was typical of this sort of reward.

Of the neurones which responded to visual stimuli that were rewarding, relatively few responded to all the rewarding stimuli used. That is, only a few (1.8%) ventral striatal neurones responded both to the sight of food and to the positive discriminative stimulus (the S +) in a visual discrimination task. Instead, the reward-related neuronal responses were typically more context or stimulus-dependent, responding, for example, to the sight of food but not to the S + which signified food (4.3%), differentially to the S + or S − but not to food (4.0%), or to food if shown in one context but not in another context. Some other neurones

(1.4%) responded to aversive stimuli. These neurones did not respond simply in relation to arousal, which was produced in control tests by inputs from different modalities, for example by touch of the leg.

These neurones with reinforcement-related responses represented 13.9% of the cells recorded in the ventral striatum. They may receive their inputs from structures such as the amygdala in which some neurones with similar responses are found (Sanghera, Rolls and Roper-Hall, 1979; Rolls, 1992b).

Thus the majority of the neurones recorded in the ventral striatum did not have unconditional sensory responses; the response typically depended on memory. The function of this part of the striatum does not appear to be purely sensory. Rather, it may provide one route for memory-related emotional and motivational stimuli to influence motor output. This is consistent with the hypothesis that the ventral striatum is a link to behaviour for learned incentive (e.g. rewarding) stimuli (see, e.g., Everitt and Robbins, 1992), and also for other limbic-processed stimuli such as faces and novel stimuli (Mogenson *et al.*, 1980; Rolls, 1984b, 1989a, 1990a; Rolls and Williams, 1987a,b; Rolls and Johnstone, 1992; Williams *et al.*, 1993). The role of the ventral striatum in feeding may thus be to provide a route for learned incentive stimuli such as the sight of food to influence behavioural responses such as approach to food.

9.8. Functions of the striatum in feeding and reward

These neurophysiological studies show that different regions of the striatum may be involved in orientation to environmental stimuli, in the use of such stimuli in the preparation for and initiation of movements, in the execution of movements, in stimulus–response connections appropriate for particular responses made in particular situations to particular stimuli, and in allowing learned reinforcers, including food, to influence behaviour. Because of its many inputs from brain systems involved in feeding and from other brain systems involved in action, it may provide a crucial route for signals which have been decoded through sensory pathways and limbic structures, and which code for the current reward value of visual, olfactory, and taste stimuli, to be interfaced in an arbitration mechanism, to produce behavioural output. The issue then arises of how the striatum, and more generally the basal ganglia, might operate in this way.

We can start to address this issue by commenting on what functions must be performed in order to link the signals that control feeding to action. It has been illustrated above how representations of objects in the environment are produced through sensory pathways. After this stage, the sensory signals are interfaced to motivational state, so that the signals in for example the orbitofrontal cortex reflect not only the taste of the food, but also whether the monkey is hungry. In these structures the signals are also passed through association memories, so that the outputs reflect whether a particular stimulus has been previously, and is still, associated with a reward. (This process will include conditioned effects from previous ingestion which reflect the energy obtained from the food, sickness produced by it, etc.) These neurones thus reflect the reward value of food, and neurones in this system can be driven by visual, olfactory, and/or taste stimuli. Now, it is of fundamental importance that these reward-related signals should not be interfaced directly to feeding movements, such as chewing. The brainstem systems which perform such functions in a limited way may assist in the execution of feeding movements later in the feeding sequence. What is required first is that the food-related reward signals should enter an arbitration mechanism, which takes into account not only the other rewards (with their magnitude) that are currently available, but also the cost of obtaining each reward.

A proposal that the basal ganglia perform this function has been developed at the conceptual level elsewhere (Rolls, 1984; Rolls and Williams, 1987a; Rolls and Johnstone, 1992). The striatum receives the appropriate signals for this function. These include not only information from limbic structures and the orbitofrontal cortex about rewards available and cognitive information from the association areas of the cerebral cortex but also information from the motor, premotor, somatosensory and parietal cortical areas about ongoing action and movements. The hypothesis is that the striatum, followed by the globus pallidus and substantia nigra, provide a two-stage system for bringing these signals together. Such convergence is implied and provided for by the dendritic organization of the basal ganglia (Percheron *et al.*, 1984; Rolls, 1984). By virtue of such anatomy, the output stages of the basal ganglia would signal that food was available and was currently rewarding, and that the costs of obtaining it were not too high. Costs might be indicated by frontal cortical inputs, combined with, for example, signals from parietal and premotor areas reflecting what position in space is being visually fixated, and the actions or movements being made in order to approach that position in space. These signals might well lead to the activation of individual neurones as a result of previous repeated co-occurrence that has induced synaptic modification in the basal ganglia. The output of such neurones would then indicate that actions should be made to approach and obtain rewarding targets in the environment. If a signal arrived in the basal ganglia from another cortical or limbic area (indicating, for example, that a novel visual stimulus had appeared), then lateral inhibitory connections within the striatum would lead to interruption of feeding, and would implement an arbitration system.

It is suggested that part of what is implemented in this way is a relatively simple interface between sensory and action systems. The signals reaching the striatum need only to signify that reward is available, with the details about where it is being implied for example by which position in space is being visually fixated and is the current subject of attention. Limiting sensory analysis in this way allows the reward signal, reflecting the output of the "what" system in perception, to be linked unambiguously with the output of the "where" system in perception (cf. Rolls, 1992a, p.18; Rolls, 1991). This is part of what simplifies the interface of the systems that specify what in the environment should be the target for action with the systems which lead to action. The action system can be organized to perform general-purpose operations, which are not different for each type of goal object or reward, in order to approach and acquire the objects in space which are currently the focus of attention. Of course, if the object that is the object of attention is associated with punishment, as shown by inputs carried by other neurones from limbic structures, then the basal ganglia should access or "look up" operations learned previously that will lead to withdrawal from what is at the current focus of attention.

The ways in which reward systems interface to action systems, and in which the basal ganglia may operate to map signals into actions, are discussed more fully elsewhere (Rolls, 1984, 1990b; Rolls and Williams, 1987a; Rolls and Johnstone, 1992). These ideas need a great deal of further development, but they address some of the crucial unanswered questions about how perceptual and motivational systems involved in decoding food reward, for example, are interfaced to the action systems that produce ingestion and other behaviour. Understanding the functions of the basal ganglia is likely to be crucial in solving these problems.

9.9. Summary and conclusions

1. There is a population of neurones in the lateral hypothalamus and adjoining substantia innominata which respond to the sight and/or taste of food if the animal is hungry. The responses of these neurones may reflect the rewarding value or pleasantness of food, for stimulation in this region can mimic the reward value of food. After satiation with one food, these hypothalamic neurones no longer respond to that food, but they may still respond to other foods which have not been eaten; this sensory-specific satiety makes variety an important determinant of human food intake. These neurones may also be on a route for forebrain-decoded stimuli, such as the sight of food, to produce autonomic responses, such as salivation and insulin release.

2. Visual information about foods reaches the hypothalamus through temporal lobe structures such as the inferior temporal visual cortex and amygdala, with the amygdala being important for learning which visual stimuli are foods. The orbitofrontal cortex contains a population of neurones which also appear to be important in correcting feeding responses as a result of learning. Its functions are closely related to those of the amygdala but, especially in primates, it may be able to implement a more rapid system for constant updating of associations between stimuli and primary reinforcers.

3. Activity in gustatory pathways up to and including the primary taste cortex of primates reflects what the taste is. After this, in the secondary taste cortex (part of the orbitofrontal cortex), activity in the gustatory system reflects motivational state, and thus the reward value of the taste. There is also an olfactory processing region in the orbitofrontal cortex where olfactory responses similarly reflect

motivational state. In this region, some convergence between olfactory and taste signals occurs; the flavour of food may therefore be computed here.

4. The striatum contains neural systems which are important for the initiation of different types of movements and behaviour, including feeding. It may be crucial in bringing together food reward signals with information from many parts of the cerebral cortex which are required to initiate actions towards foods.

References

Aggleton, J. P. and Passingham, R. E. (1981). Syndrome produced by lesions of the amygdala in monkeys (*Macaca mulatta*). *Journal of Comparative and Physiological Psychology*, **95,** 961 –977.

Aggleton, J. P. and Passingham, R. E. (1982). An assessment of the reinforcing properties of foods after amygdaloid lesions in rhesus monkeys. *Journal of Comparative and Physiological Psychology.*, **96,** 71–77.

Akert, K., Gruesen, R. A., Woolsey, C. N. and Meyer, D. R. (1961). Kluver-Bucy syndrome in monkeys with neocortical ablations of temporal lobe. *Brain*, **84,** 480–498.

Amaral, D. G., Price, J. L., Pitkanen, A. and Carmichael, S. T. (1992). Anatomical organization of the primate amygdaloid complex. J. P. Aggleton, (Ed.) *The Amygdala* (pp. 1–66). Wiley-Liss: New York.

Anand, B. K., Brobeck, J. R. (1951). Localization of a feeding center in the hypothalamus of the rat. *Proceedings of Society for Experimental Biology and Medicine*, **77,** 323–324.

Aou, S., Oomura, Y., Lenard, L., Noshino, H., Inokuchi, A., Minami, T. and Misaki, H. (1984). Behavioral significance of monkey hypothalamic glucose-sensitive neurons. *Brain Res.*, **302,** 69–74.

Baylis, L. L. and Gaffan, D. (1991). Amygdalectomy and ventromedial prefrontal ablation produce similar deficits in food choice and in simple object discrimination learning for an unseen reward. *Exp. Brain Res.*, **86,** 617–622.

Baylis, L. L. and Rolls, E. T. (1991). Responses of neurons in the primate taste cortex to glutamate. *Physiology and Behavior*, **49,** 973–979.

Beckstead, R. M. and Norgren, R. (1979). An autoradiographic examination of the central distribution of the trigeminal, facial, glossopharyngeal, and vagal nerves in the monkey. *J. Comp. Neurol.*, **184,** 455–472.

Beckstead, R. M., Morse, J. R. and Norgren, R. (1980). The nucleus of the solitary tract in the monkey: projections to the thalamus and brainstem nuclei. *J. Comp. Neurol.*, **190,** 259–282.

Booth, D. A. and Toase, A.-M. (1983). Conditioning of hunger/satiety signals as well as flavour cues in dieters. *Appetite*, **4,** 235–236.

Borsini, F. and Rolls, E. T. (1984). Role of noradrenaline and serotonin in the basolateral region of the amygdala in food preferences and learned taste aversions in the rat. *Physiology and Behavior*, **33,** 37–43.

Burton, M. J., Rolls, E. T. and Mora, F. (1976). Effects of hunger on the responses of neurons in the lateral hypothalamus to the sight and taste of food. *Exptl. Neurol.*, **51,** 668–77.

Butter, C. M. (1969). Perseveration in extinction and in discrimination reversal tasks following selective prefrontal ablations in *Macaca mulatta. Physiol. Behav.*, **4,** 163–71.

Butter, C. M., McDonald, J. A. and Snyder, D. R. (1969). Orality, preference behavior, and reinforcement value of non-food objects in monkeys with orbital frontal lesions. *Science*, **164,** 1306–1307.

Caan, W., Perrett, D. I. and Rolls, E. T. (1984). Responses of striatal neurons in the behaving monkey. 2. Visual processing in the caudal neostriatum. *Brain Research*, **290,** 53–65.

Cabanac, M. (1971). Physiological role of pleasure. *Science*, **173,** 1103–1107.

Cabanac, M. and Duclaux, R. (1970). Specificity of internal signals in producing satiety for taste stimuli. *Nature*, **227,** 966–967.

Cabanac, M. and Fantino, M. (1977). Origin of olfacto-gustatory alliesthesia: intestinal sensitivity to carbohydrate concentration? *Physiol. Behav.*, **10,** 1039–1045.

Clark, J. M., Clark, A. J. M., Bartle, A. and Winn, P. (1991). The regulation of feeding and drinking in rats with lesions of the lateral hypothalamus made by N-methyl-D-aspartate. *Neuroscience*, **45**, 631–640.

Divac, I. (1975). Magnocellular nuclei of the basal forebrain project to neocortex, brain stem, and olfactory bulb. Review of some functional correlates. *Brain Res.*, **93**, 385–98.

Dunn, L. T. and Everitt, B. J. (1988). Double dissociations of the effects of amygdala and insular cortex lesions on conditioned taste aversion, passive avoidance, and neophobia in the rat using the excitotoxin ibotenic acid. *Behav. Neurosci.*, **102**, 3–23.

Dunnett, S. B., Lane, D. M. and Winn, P. (1985). Ibotenic acid lesions of the lateral hypothalamus: comparison with 6-hydroxydopamine-induced sensorimotor deficits. *Neuroscience*, **14**, 509–518.

Ewart, W. (1993). Hepatic and other parenteral visceral afferents affecting ingestive behaviour. In D. A. Booth (Ed.), *Neurophysiology of Ingestion*. Pergamon, Oxford.

Everitt, B. J. and Robbins, T. W. (1992). Amygdala-ventral striatal interactions and reward-related processes. In J. P. Aggleton (Ed.), *The Amygdala* (pp. 401–430). Wiley: Chichester.

Fuster, J. M. (1989). *The Prefrontal Cortex*, 2nd edition. Raven Press, New York.

Gaffan, D. (1992). Amygdala and the memory of reward. In J. P. Aggleton (Ed.), *The Amygdala* (pp. 471–483). Wiley-Liss: New York.

Gaffan, D. and Harrison, S. (1987). Amygdalectomy and disconnection in visual learning for auditory secondary reinforcement by monkeys. *J. Neurosci.*, **7**, 2285–2292.

Gaffan, E. A., Gaffan, D. and Harrison, S. (1988). Disconnection of the amygdala from visual association cortex impairs visual reward-association learning in monkeys. *J. Neurosci.*, **8**, 3144–3150.

Gaffan, D., Gaffan, E. A. and Harrison, S. (1989). Visual-visual associative learning and reward-association learning in monkeys: the role for the amygdala. *J. Neurosci.*, **9**, 558–564.

Gallistel, C. R. Beagley, G. (1971). Specificity of brain-stimulation reward in the rat. *J. Comp. Physiol. Psychol.*, **76**, 199–205.

Gibbs, J., Maddison, S. P. and Rolls, E. T. (1981). The satiety role of the small intestine in sham feeding rhesus monkeys. *J. Comp. Physiol. Psychol.*, **95**, 1003–1015.

Gibson, E. L. and Booth, D. A. (1989). Dependence of carbohydrate-conditioned flavor preference on internal state in rats. *Learning and Motivation*, **20**, 36–47.

Grill, H. J. and Norgren, R. (1978). Chronically decerebrate rats demonstrate satiation but not bait shyness. *Science*, **201**, 267–269.

Groenewegen, H. J., Berendse, H. W., Meredith, G. E., Haber, S. N., Voorn, P., Wolters, J. G. and Lohman, A. H. M. (1991). Functional anatomy of the ventral, limbic system-innervated striatum. In P. Willner and J. Scheel-Kruger (Eds.), *The Mesolimbic Dopamine System: From Motivation to Action* (pp. 19–60). Wiley: Chichester.

Gross, C. G., Bender, D. B. and Gerstein, G. L. (1979). Activity of inferior temporal neurons in behaving monkeys. *Neuropsychologia*, **17**, 215–229.

Grossman, S. P. (1967). *A Textbook of Physiological Psychology*. Wiley: New York.

Grossman, S. P. (1973). *Essentials of Physiological Psychology*. Wiley: New York.

Heimer, L., Switzer, R. D. and Van Hoesen, G. W. (1982). Ventral striatum and ventral pallidum. Additional components of the motor system? *Trends Neurosci.*, **5**, 83–87.

Heimer, L. and Alheid, G. F. (1990). Piecing together the puzzle of basal forebrain anatomy. In T. C. Napier, P. W. Kalivas and I. Hanin (Eds.), *The Basal Forebrain: Anatomy to Function*. Plenum: New York.

Herzog, A. G. and Van Hoesen, G. W. (1976). Temporal neocortical afferent connections to the amygdala in the rhesus monkey. *Brain Res.*, **115**, 57–69.

Hoebel, B. G. (1969). Feeding and self-stimulation. *Ann. N. Y. Acad. Sci.*, **157**, 757–778.

Iversen, S. D. and Mishkin, M. (1970). Preserverative interference in monkey following selective lesions of the inferior prefrontal convexity. *Exp. Brain Res.*, **11**, 376–386.

Jarvis, C. D. and Mishkin, M. (1977). Responses of cells in the inferior temporal cortex of monkeys during visual discrimination reversals. *Soc. Neurosci. Abstr.*, **3**, 1794.

Jones, B. and Mishkin, M. (1972). Limbic lesions and the problem of stimulus-reinforcement associations. *Exptl. Neurol.*, **36**, 362–377.

Kievit, J. and Kuypers, H. G. J. M. (1975). Subcortical afferents to the frontal lobe in the rhesus monkey studied by means of retrograde horseradish peroxidase transport. *Brain Res.*, **85**, 262–266.

Kluver, H. and Bucy, P. C. (1939). Preliminary analysis of functions of the temporal lobes in monkeys. *Archives of Neurology and Psychiatry*, **42**, 979–1000.

LeDoux, J. E., Iwata, J., Cicchetti, P. and Reis, D. J. (1988). Different projections of the central amygdaloid nucleus mediate autonomic and behavioral correlates of conditioned fear. *J. Neurosci.*, **8**, 2517–2529.

Le Magnen, J. (1992). *Neurobiology of Feeding and Nutrition*. Academic Press: San Diego.

Marshall, J. F., Richardson, J. S. and Teitelbaum, P. (1974). Nigrostriatal bundle damage and the lateral hypothalamic syndrome. *J. Comp. Physiol. Psychol.*, **87**, 808–830.

Mei, N. (1993). Gastrointestinal chemoreception and its behavioural role. In D. A. Booth (Ed.), *Neurophysiology of Ingestion*. Pergamon, Oxford.

Mogenson, G. J., Jones, D. L. and Yim, C. Y. (1980). From motivation to action: functional interface between the limbic system and the motor system. *Progr. Neurobiol.*, **14**, 69–97.

Mora, F., Rolls, E. T. and Burton, M. J. (1976). Modulation during learning of the responses of neurons in the hypothalamus to the sight of food. *Exptl. Neurol.*, **53**, 508–519.

Mora, F., Avrith, D. B., Phillips, A. G. and Rolls, E. T. (1979). Effects of satiety on self-stimulation of the orbitofrontal cortex in the monkey. *Neuroscience Letters*, **13**, 141–145.

Nauta, W. J. H. (1961). Fiber degeneration following lesions of the amygdaloid complex in the monkey. *J. Anat.*, **95**, 515–531.

Nauta, W. J. H. and Domesick, V. B. (1978). Crossroads of limbic and striatal circuitry: Hypothalamonigral connections. In K. E. Livingston and O. Hornykiewicz (Eds.), *Limbic Mechanisms* (pp. 75–93). New York: Plenum.

Nishijo, H., Ono, T. and Nishino, H. (1988). Single neuron responses in amygdala of alert monkey during complex sensory stimulation with affective significance. *J. Neurosci.*, **8**, 3570–3583.

Olds, J. (1977). *Drives and Reinforcements: Behavioral Studies of Hypothalamic Functions*. New York: Raven Press.

Ono, T., Nishino, H., Sasaki, K., Fukuda, M. and Muramoto, K. (1980). Role of the lateral hypothalamus and amygdala in feeding behavior. *Brain Res. Bull.*, **5**, Suppl. 4, 143–149.

Ono, T., Tamura, R., Nishijo, H., Nakamura, K. Tabuchi, E. (1989). Contribution of amygdala and LH neurons to the visual information processing of food and non-food in the monkey. *Physiol. Behav.*, **45**, 411–421.

Ono, T. and Nishijo, H. (1992). Neurophysiological basis of the Kluver-Bucy syndrome: responses of monkey amygdaloid neurons to biologically significant objects. In J. P. Aggleton (Ed.), *The Amygdala* (pp. 167–190). Wiley-Liss: New York.

Oomura, Y. and Yoshimatsu, H. (1984). Neural network of glucose monitoring system. *Journal of the Autonomic Nervous System*, **10**, 359–372.

Percheron, G., Yelnik, J. and Francois, C. (1984b). The primate striato-pallido-nigral system: an integrative system for cortical information. In J. S. McKenzie, R. E. Kemm and L. N. Wilcox (Eds.), *The Basal Ganglia: Structure and Function* (pp. 87–105). Plenum, New York.

Ridley, R. M., Hester, N. S. and Ettlinger, G. (1977). Stimulus- and response-dependent units from the occipital and temporal lobes of the unanaesthetized monkey performing learnt visual tasks. *Exp. Brain Res.*, **27**, 539–552.

Ritter, S. (1986). Glucoprivation and the glucoprivic control of food intake. In R. C. Ritter, S. Ritter and C. D. Barnes (Eds.), *Feeding Behavior: Neural and Humoral Controls* (pp. 271–313). Academic Press: New York.

Robbins, T. W. and Everitt, B. J. (1992). Functions of dopamine in the dorsal and ventral striatum. *Seminars in the Neurosciences*, **4**, 119–128.

Rolls, B. J. (1990). The role of sensory-specific satiety in food intake and food selection. In E. D. Capaldi and T. L. Powley (Eds.), *Taste, Experience, and Feeding* (pp. 197–209). American Psychological Association: Washington, D. C.

Rolls, B. J. and Rolls, E. T. (1973). Effects of lesions in the basolateral amygdala on fluid intake in the rat. *Journal of Comparative and Physiological Psychology*, **83**, 240–247.

Rolls, B. J., Rolls, E. T., Rowe, E. A. and Sweeney, K. (1981). Sensory-specific satiety in man. *Physiol. Behav.*, **27**, 137–142.

Rolls, B. J., Rowe, E. A., Rolls, E. T., Kingston, B., Megson, A. and Gunary, R. (1981). Variety in a meal enhances food intake in man. *Physiol. Behav.*, **26**, 215–221.

Rolls, B. J., Rowe, E. A. and Rolls, E. T. (1982). How sensory properties of foods affect human feeding behavior. *Physiol. Behav.*, **29**, 409–417.

Rolls, B. J., Van Duijenvoorde, P. M. and Rowe, E. A. (1983). Variety in the diet enhances intake in a meal and contributes to the development of obesity in the rat. *Physiol. Behav.*, **31**, 21–27.

Rolls, B. J., Van Duijenvoorde, P. M. and Rolls, E. T. (1984). Pleasantness changes and food intake in a varied four course meal. *Appetite*, **5**, 337–348.

Rolls, B. J. and Hetherington, M. (1989). The role of variety in eating and body weight regulation. In R. Shepherd (Ed.), *Handbook of the Psychophysiology of Human Eating* (pp. 57–84). Wiley: Chichester.

Rolls, E. T. (1975). *The Brain and Reward*. Oxford: Pergamon.

Rolls, E. T. (1976). The neurophysiological basis of brain-stimulation reward. In A. Wauquier and E. T. Rolls, (Eds.), *Brain-stimulation Reward* (pp. 65–87)Amsterdam: North-Holland.

Rolls, E. T. (1979). Effects of electrical stimulation of the brain on behavior. In K. Connolly (Ed.), *Psychology Surveys*, Vol. 2 (pp. 151–169). Allen and Unwin: Hemel Hempstead, U.K.

Rolls, E. T. (1981a). Processing beyond the inferior temporal visual cortex related to feeding, memory,and striatal function. In Y. Katsuki, R. Norgren and M. Sato, (Eds.), *Brain Mechanisms of Sensation* (pp. 241–269). Wiley, New York.

Rolls, E. T. (1981b). Central nervous mechanisms related to feeding and appetite. *Brit. Med. Bull.*, **37**, 131–134.

Rolls, E. T. (1982). Feeding and reward. In B. G. Hoebel and D. Novin (Eds.), *The Neural Basis of Feeding and Reward* (pp. 323–337). Brunswick, Maine: Haer Institute for Electrophysiological Research.

Rolls, E. T. (1984). Activity of neurons in different regions of the striatum of the monkey. In J. S. McKenzie, R. E. Kemm and L. N. Wilcox (Eds.), *Basal Ganglia: Structure and Function* (pp. 467–493). New York: Plenum.

Rolls, E. T. (1986). Neuronal activity related to the control of feeding. In R. C. Ritter, S. Ritter and C. D. Barnes (Eds.), *Feeding Behavior: Neural and Humoral Controls* (pp. 163–190). Academic Press: New York.

Rolls, E. T. (1987). Information representation, processing and storage in the brain: analysis at the single neuron level. In J. -P. Changeux and M. Konishi (Eds.), *The Neural and Molecular Bases of Learning* (pp. 503–540). Wiley: Chichester.

Rolls, E. T. (1989). Information processing in the taste system of primates. *Journal of Experimental Biology*, **146**, 141–164.

Rolls, E. T. (1990a). Functions of different regions of the basal ganglia. In G. M. Stern (Ed.), *Parkinson's Disease* (pp. 151–184). Chapman and Hall: London.

Rolls, E. T. (1990b). A theory of emotion, and its application to understanding the neural basis of emotion. *Cognition and Emotion*, **4**, 161–190.

Rolls, E. T. (1991). Neural organisation of higher visual functions. *Current Opinion in Neurobiology*, **1**, 274–278.

Rolls, E. T. (1992a). Neurophysiological mechanisms underlying face processing within and beyond the temporal cortical visual areas. *Philosophical Transactions of the Royal Society*, **335**, 11–21.

Rolls, E. T. (1992b). Neurophysiology and functions of the primate amygdala. In J. P. Aggleton (Ed.), *The Amygdala* (pp. 143–165). Wiley-Liss: New York.

Rolls, E. T. and Rolls, B. J. (1973). Altered food preferences after lesions in the baso-lateral region of the amygdala in the rat. *Journal of Comparative and Physiological Psychology*, **83**, 248–259.

Rolls, E. T., Burton, M. J. and Mora, F. (1976). Hypothalamic neuronal responses associated with the sight of food, *Brain Res.*, **111**, 53–66.

Rolls, E. T., Judge, S. J. Sanghera, M. K. (1977). Activity of neurons in the inferotemporal cortex of the alert monkey. *Brain Res.*, **130**, 229–238.

Rolls, E. T. and Rolls, B. J. (1977). Activity of neurones in sensory, hypothalamic, and motor areas during feeding in the monkey. In Y. Katsuki, M. Sato, S. F. Takagi and Y. Oomura

(Eds.), *Food Intake and Chemical Senses* (pp. 525–549). Tokyo: University of Tokyo Press.

Rolls, E. T., Sanghera, M. K. and Roper-Hall, A. (1979). The latency of activation of neurons in the lateral hypothalamus and substantia innominata during feeding in the monkey. *Brain Res.*, **164**, 121–135.

Rolls, E. T., Burton, M. J. and Mora, F. (1980). Neurophysiological analysis of brain-stimulation reward in the monkey. *Brain Research*, **194**, 339–357.

Rolls, E. T. and Rolls, B. J. (1982). Brain mechanisms involved in feeding. In L. M. Barker (Ed.), *Psychobiology of Human Food Selection* (pp. 33–62). Westport, Conn.: AVI Publishing Co.

Rolls, E. T., Rolls, B. J. and Rowe, E. A. (1983). Sensory-specific and motivation-specific satiety for the sight and taste of food and water in man. *Physiology and Behavior*, **30**, 185–192.

Rolls, E. T., Thorpe, S. J. and Maddison, S. P. (1983). Responses of striatal neurons in the behaving monkey. 1. Head of the caudate nucleus. *Behavioural Brain Research*, **7**, 179–210.

Rolls, E. T., Thorpe, S. J., Boytim, M., Szabo, I. and Perrett, D. I. (1984). Responses of striatal neurons in the behaving monkey. 3. Effects of iontophoretically applied dopamine on normal responsiveness. *Neuroscience*, **12**, 1202–1212.

Rolls, E. T. and de Waal, A. W. L. (1985). Long-term sensory-specific satiety: evidence from an Ethiopian refugee camp. *Physiology and Behavior*, **34**, 1017–1020.

Rolls, E. T. and Williams, G. V. (1987a). Sensory and movement-related neuronal activity in different regions of the primate striatum. In J. S. Schneider and T. I. Lidsky (Eds.), *Basal Ganglia and Behavior: Sensory Aspects and Motor Functioning* (pp. 37–59). Hans Huber: Bern.

Rolls, E. T. and Williams, G. V. (1987b). Neuronal activity in the ventral striatum of the primate. In M. B. Carpenter and A. Jayamaran (Eds.), *The Basal Ganglia II—Structure and Function—Current Concepts* (pp. 349–356). Plenum: New York.

Rolls, E. T., Murzi, E., Yaxley, S., Thorpe, S. J., and Simpson, S. J. (1986). Sensory-specific satiety: food-specific reduction in responsiveness of ventral forebrain neurons after feeding in the monkey. *Brain Research*, **368**, 79–86.

Rolls, E. T., Scott, T. R., Sienkiewicz, Z. J. and Yaxley, S. (1988). The responsiveness of neurones in the frontal opercular gustatory cortex of the macaque monkey is independent of hunger. *Journal of Physiology*, **397**, 1–12.

Rolls, E. T., Sienkiewicz, Z. J. and Yaxley, S. (1989). Hunger modulates the responses to gustatory stimuli of single neurons in the caudolateral orbitofrontal cortex of the macaque monkey. *European Journal of Neuroscience*, **1**, 53–60.

Rolls, E. T. and Treves, A. (1990). The relative advantages of sparse versus distributed encoding for associative neuronal networks in the brain. *Network*, **1**, 407–421.

Rolls, E. T., Yaxley, S. and Sienkiewicz, Z. J. (1990). Gustatory responses of single neurons in the orbitofrontal cortex of the macaque monkey. *Journal of Neurophysiology*, **64**, 1055–1066.

Rolls, E. T. and Mason, R. (1991). Convergence of olfactory and taste inputs in the primate orbitofrontal cortex: modification by learning. *European Journal of Neuroscience*, Supplement **4**, 88.

Rolls, E. T. and Johnstone, S. (1992). Neurophysiological analysis of striatal function. In G. Vallar, S. F. Cappa and C. Wallesch (Eds.), *Neuropsychological Disorders Associated with Subcortical Lesions* (pp. 61–97). Oxford University Press: Oxford.

Rosenkilde, C. E. (1979). Functional heterogeneity of the prefrontal cortex in the monkey: a review. *Behav. Neural Biol.*, **25**, 301–345.

Russchen, F. T., Amaral, D. G. and Price, J. L. (1985). The afferent connections of the substantia innominata in the monkey, *Macaca fascicularis. J. Comp. Neurol.*, **242**, 1–27.

Sanghera, M. K., Rolls, E. T. and Roper-Hall, A. (1979). Visual responses of neurons in the dorsolateral amygdala of the alert monkey. *Exptl. Neurol.*, **63**, 610–626.

Saper, C. B., Loewy, A. D., Swanson, L. W. and Cowan, W. M. (1976). Direct hypothalamo-autonomic connections. *Brain Res.*, **117**, 305–312.

Saper, C. B., Swanson, L. W. and Cowan, W. M. (1979). An autoradiographic study of the efferent connections of the lateral hypothalamic area in the rat. *J. Comp. Neurol.*, **183**, 689–706.

Sato, T., Kawamura, T., and Iwai, E. (1980). Responsiveness of inferotemporal single units to visual pattern stimuli in monkeys performing discrimination. *Exp. Brain Res.*, **38**, 313–319.

Scott, T. R., Yaxley, S., Sienkiewicz, Z. J. and Rolls, E. T. (1986). Taste responses in the nucleus tractus solitarius of the behaving monkey. *Journal of Neurophysiology*, **55**, 182–200.

Scott, T. R., Yaxley, S., Sienkiewicz, Z. J. and Rolls, E. T. (1986). Gustatory responses in the frontal opercular cortex of the alert cynomolgus monkey. *Journal of Neurophysiology*, **56**, 876–890.

Scott, T. R. and Giza, B. K. (1992). Gustatory control of ingestion. In D. A. Booth (Ed.), *Neurophysiology of Ingestion*. Pergamon: Oxford.

Spiegler, B. J. and Mishkin, M. (1981). Evidence for the sequential participation of inferior temporal cortex and amygdala in the acquisition of stimulus-reward associations. *Behav. Brain Res.*, **3**, 303–317.

Stellar, E. (1954). The physiology of motivation. *Psychol. Rev.*, **61**, 5–22.

Stricker, E. M. (1984). Brain monoamines and the control of food intake. *Int. J. Obes.*, **8**, (Suppl. 1), in press.

Stricker, E. M. and Zigmond, M. J. (1976). Recovery of function after damage to central catecholamine-containing neurons: a neurochemical model for the lateral hypothalamic syndrome. *Prog. Psychobiol. Physiol. Psychol.*, **6**, 121–188.

Swash, M. (1989). John Hughlings Jackson: a historical introduction. In C. Kennard and M. Swash (Eds.), *Hierarchies in Neurology: A Reappraisal of a Jacksonian Concept.* (pp. 3–10). Springer: London.

Tanabe, T., Yarita, H., Iino, M. Ooshima, Y. and Takagi, S. F. (1975a). An olfactory projection area in orbitofrontal cortex of the monkey. *J. Neurophysiol.*, **38**, 1269–1283.

Tanabe, T., Iino, M. and Takagi, S. F. (1975b). Discrimination of odors in olfactory bulb, pyriform-amygdaloid areas, and orbitofrontal cortex of the monkey. *J. Neurophysiol.*, **38**, 1284–1296.

Tanaka, D. (1973). Effects of selective prefrontal decortication on escape behavior in the monkey. *Brain Res.*, **53**, 161–173.

Thorpe, S. J., Rolls, E. T. and Maddison, S. (1983). Neuronal activity in the orbitofrontal cortex of the behaving monkey. *Experimental Brain Research*, **49**, 93–115.

Ungerstedt, U. (1971). Adipsia and aphagia after 6-hydroxydopamine induced degeneration of the nigrostriatal dopamine system. *Acta Physiol. Scand.*, **81**, (Suppl. 367), 95–122.

Weiskrantz, L. (1956). Behavioral changes associated with ablation of the amygdaloid complex in monkeys. *J. Comp. Physiol. Psychol.*, **49**, 381–391.

Weiskrantz, L. and Saunders, R. C. (1984). Impairments of visual object transforms in monkeys. *Brain*, **107**, 1033–1072.

Wiggins, L. L., Rolls, E. T. and Bayliss, G. C. (1987). Afferent connections of the orbitofrontal cortex taste area of the primate. *Chemical Senses*, **12**, 206.

Williams, G. V., Rolls, E. T., Leonard, C. M. and Stern, C. (1993). Neuronal responses in the ventral striatum of the behaving monkey. *Behavioural Brain Research*, **55**, 243–252.

Wilson, F. A. W. and Rolls, E. T. (1990). Neuronal responses related to reinforcement in the primate basal forebrain. *Brain Research*, **502**, 213–231.

Wilson, F. A. W. and Rolls, E. T. (1993). The primate amygdala and reinforcement: a dissociation between rule-based and associatively-mediated memory revealed in amygdala neuronal activity.

Winn, P., Tarbuck, A. and Dunnett, S. B. (1984). Ibotenic acid lesions of the lateral hypothalamus: comparison with electrolytic lesion syndrome. *Neuroscience*, **12**, 225–240.

Winn, P., Clark, A., Hastings, M., Clark, J., Latimer, M., Rugg, E. and Brownlee, B. (1990). Excitotoxic lesions of the lateral hypothalamus made by N-methyl-D-aspartate in the rat: behavioural, histological and biochemical analyses. *Exp. Brain Res.*, **82**, 628–636.

Yaxley, S., Rolls, E. T., Sienkiewicz, Z. J. and Scott, T. R. (1985). Satiety does not affect gustatory activity in the nucleus of the solitary tract of the alert monkey. *Brain Research*, **347**, 85–93.

Yaxley, S., Rolls, E. T. and Sienkiewicz, Z. J. (1988). The responsiveness of neurones in the insular gustatory cortex of the macaque monkey is independent of hunger. *Physiology and Behavior*, **42**, 223–229.

Yaxley, S., Rolls, E. T. and Sienkiewicz, Z. J. (1990). Gustatory responses of single neurons in the insula of the macaque monkey. *Journal of Neurophysiology*, **63**, 689–700.

Subject Index